Sergio Ayvar-Serna
José Francisco Díaz-Nájera
Antonio Mena-Bahena

INTEGRATED IN vitro CONTROL OF Pestalotiopsis microspora Speg.

Sergio Ayvar-Serna
José Francisco Díaz-Nájera
Antonio Mena-Bahena

INTEGRATED IN vitro CONTROL OF Pestalotiopsis microspora Speg.

Guava clove management

ScienciaScripts

Imprint

Cover image: www.ingimage.com

This book is a translation from the original published under ISBN 978-613-9-40439-1.

Publisher:
Sciencia Scripts
is a trademark of
Dodo Books Indian Ocean Ltd. and OmniScriptum S.R.L publishing group

120 High Road, East Finchley, London, N2 9ED, United Kingdom
Str. Armeneasca 28/1, office 1, Chisinau MD-2012, Republic of Moldova, Europe
Printed at: see last page
ISBN: 978-620-7-74403-9

CONTENT

I.INTRODUCTION

Guava plantations established in the Mexican Republic are affected by the endemic disease known as "clove" caused by the fungus Pestalotia microspora Speg., which infects leaves, branches, flowers and fruits, in the latter it causes more damage because it detracts from their presentation, quality and commercial value, which affects the profitability of this fruit activity. The damage in The fruit appears as small brown lesions, circular in shape with a corky consistency, with the appearance of a tack or rusty nail. (Figure 1). The best alternative to fight this disease is driving integrated using jointly cultural, biological, chemical and other practices that contribute to counteract the damage. Without embargo, he preferred method farmers is the application of synthetic fungicides, because they give short-term results, they are available in different prices help increase the quantity and quality of the product; However, this control method does not guarantee the food safety of the fruit; because toxic substances are used that also contaminate the environment, generate residual waste, induce pathogen resistance and cause carcinogenic diseases in humans; For this reason, it has been decided to look for other natural alternatives compatible with agricultural production systems, among which biological control stands out through the use of beneficial microorganisms such as Trichoderma spp. which acts through direct competition, mycoparasitism and antibiosis, against various species of phytopathogenic fungi, such as P.microspora. In addition, Trichoderma presents indirect mechanisms to elicit or induce the physiological and biochemical defense of the plant, through the detoxification of toxins. and deactivation of enzymes in the infection processes; It also promotes the solubilization of nutritional elements such as phosphorus that, in their original form, are not accessible to plants; in addition,has ability of create a atmosphere favorable for root development, which increases tolerance of the plant to the stress from drought and/or nutrition (Infant et al. , 2009). In current organic agriculture, interest has increased in he exploitation of antimicrobial compounds produced by some plants such as cinnamon (Cinnamomum zeylanicum L.), garlic (Allium sativum L.),neem (Azadirachta indica Juss), and many others plant species, which present essential oils, polyphenols, tannins, saponins, anthraquinones, flavonoids, terpenes, lectins, polypeptides, phenols and many more compounds, which have antifungal activity, with the advantages of being compatible

with nature and harmless to consumers. Terpenes are mainly responsible for the antimicrobial activity of essential oils; his antimicrobial effect is based on the fact that they damage biomembranes, and interfere with vital processes such as osmosis, the synthesis of sterols and phospholipids in the cell fungosa (Montes , 2006).Given the need to strengthen the integrated control of P.microspora , interest arose in evaluating the biological effectiveness, in vitro, from Trichoderma spp ., plant extracts and chemical fungicides for select products with greater potential for check to the pathogen.

II. PROBLEM OF INVESTIGATION

In Tetipac, Gro., guava plantations suffer the attack of the endemic disease caused by the fungus P. microspora , which affects various parts of the plant in different phenological stages; but the most significant damage is in the fruit (Figure 1), where it induces the formation of lesions or necrotic scabs that mainly affect the quality of the product, which decreases the profitability of the crop.

In the study region, there is no history of formal research done to prove the etiology and alternatives to reduce damage from this problematic phytosanitary. For this reason, the interest arose in testing, under laboratory conditions, the effect of various biological, organic and chemical products, to determine the most effective ones that can be used. in the integrated management of this disease.

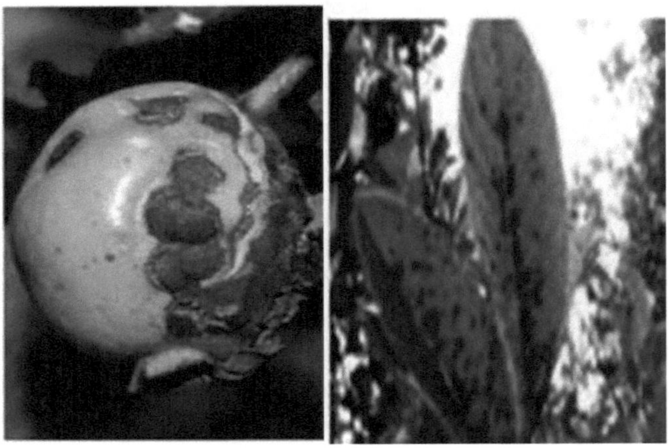

Figure 1. Damage provoked by P. microspore in fruits and sheet (Flowers, 2012).

III. GOALS AND HYPOTHESIS

By he interest of select new alternatives for he control integrated in vitro of P.microspora, HE he took to cape he present study with base in the following:

3.1. Goals

- Isolate and purify he fungus that cause the disease of nail in he crop of guava.
- Identify morphological and molecularly to the causal agent of this sickness.
- Assess pathogenicity of the species mushroom isolated.
- Know the effectiveness biological in vitro of six fungicides chemicals, six strains of Trichoderma spp . and six extracts vegetables against he fungus Q. microspore.

3.2. Hypothesis

- The species of the fungus isolated is capable of reproducing the symptoms of fruit disease and healthy leaves inoculated in vivo.
- The species from Trichoderma are effective for inhibit growth of the fungus.
- The fungicides systemic has a greater effectiveness that the contact products for inhibit development of the pathogen in in vitro conditions .
- The extracts vegetables are able of counter he growth mycelial of the fungus.

IV. REVISION OF LITERATURE

4.1. Aspects general of the crop

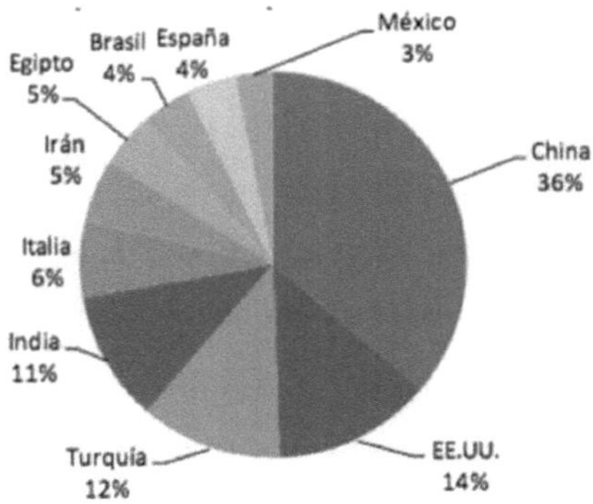

The guava is native to the tropical regions of America; from where it has spread to the whole world. Currently, it is produced mainly in the countries indicated in Figure 2, where China stands out for contributing 36% of the total world volume. Mexico participates with only 3%, and the states that lead production are Michoacán and Aguascalientes (Figure 3). In Guerrero the The largest area dedicated to this fruit tree is found in the municipality of Tetipac (Figure 4), where infection with the aerial disease known as scab or guava clove, caused by the fungus P. microspora Speg., which affects the commercial value of the product. In the present investigation, P. microspora and biological, organic and chemical control methods are the object of study. As it is the first report of this species of fungus in guava, all the information about P. microspora is presented , because there are not many publications about the first of the mentioned fungi .

4.2. Anthracnose or guava clove (P. microspora Speg)

Importance. In field it affects the appearance and

Figure 2 . Countries main producers (SAGARPA, 2014).

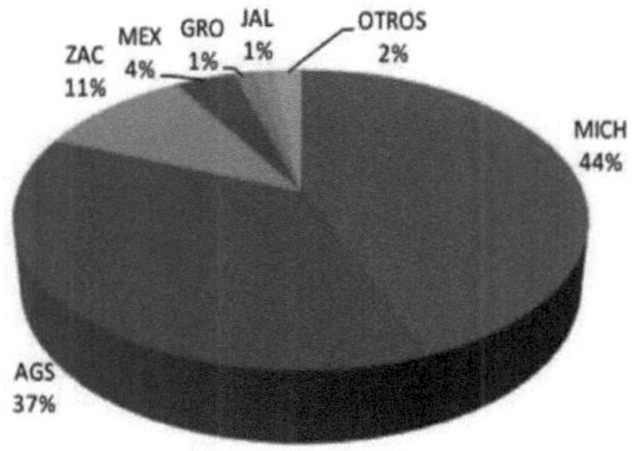

Figure 3 . state main producers (SAGARPA, 2014).

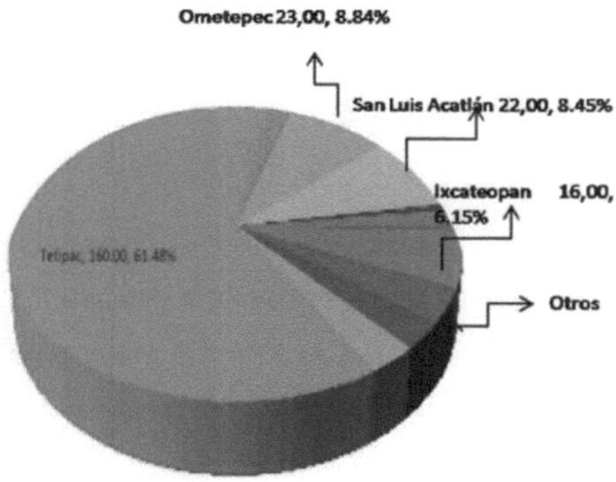

Figure 4 . Municipalities main producers (SAGARPA, 2014). quality commercial of the fruit and in postharvest causes decay dry and mummification of the product (Agríos,2005).

Distribution. It is found in Mexico, Bolivia, India, Australia, Malaysia, Mozambique, Zambia, Nigeria, Venezuela, Ecuador, Ivory Coast, Taiwan, Zimbabwe, Tanzania, Puerto Rico and many other countries (Domingo et al ., 2006).

Hosts. It infects Psidium guajava L., Acca sellowiana Berg., Vitis vinifera L. , Feijoa sellowiana Berg . and Musa paradisiaca L. (Jiménez et al., 1992).

Symptoms. The disease manifests itself on the epicarp of the young fruit, as circular, raised spots, with a corky consistency and a brown color (Figure 5), which affect the normal development of the fruit; or, it becomes covered with irregular brown scabs that, in severe attacks, cause cracks. and at the end the mummification. The leaves have stains of color coffee

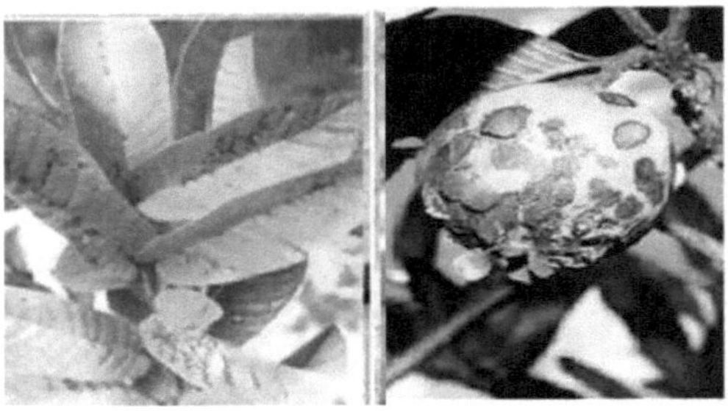

Figure 5. Symptoms of Q. microspore in leaves and fruit (Flores, 2012). shiny both on the edges and in its central part (Figure 5). Likewise, tender shoots become deformed, dry, darken and fall (Montiel et al., 2001).

Diagnosis. In the field, the preliminary diagnosis is made through the observation of the symptoms that characterize the presence of the disease (Figures 1, 5 and 11). In the laboratory, diseased tissue is placed in a humid chamber to induce the formation of typical acervules of the fungus and observe the typical conidia (Figures 6 and 21) that allow the identification of the genus of this fungus.

Taxonomy. The taxonomic position of the fungus is: Kingdom Fungi, Phylum Deuteromycota, Class Deuteromycetes, Order Melanconiales, Family Melanconiaceae, Genus Pestalotia , Species microspora (Romero, 1988; Insuasty et al., 2005).

Morphology. The fungus presents hyaline, irregularly branched, septate, smooth and short conidiophores (Figure 6); as well as dark, straight or slightly curved, ellipsoidal or fusiform conidia that have six cells, with the basal and terminal cells pointed and hyaline, and the latter having two or more hyaline apical appendages (Romero, 1988; Insuasty et al., 2005).

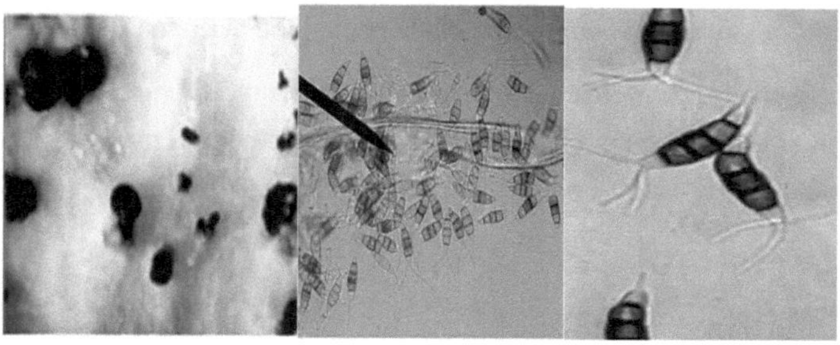

Figure 6. Acervuli (10X) and conids (40X and 100X) of Q. microspore (Insuasty et al., 2005).

Cycle biological. He fungus produces acervuli where HE form the conidia that They are easily spread by wind and splashing rainwater. The presence of periods of high relative humidity and/or rain during dry weather favors the development and reproduction of the pathogen (Figure 6); it has a saprophytic phase that helps it survive in dead tissues for prolonged periods. Penetration into the host tissues is carried out through wounds or mechanical damage (Insuasty et al., 2005; Anonymous, 2015a).

Epidemiology . The incidence and severity are related to the virulence of the pathogen, as well as the planting densities, favorable environmental conditions of high relative humidity, especially favored by excess plant foliage and. also the highest altitude. On the other hand, there is a presumption that this fungus is associated with the attack of sucking or scraping insects (Arevalo et al., 2012). The infection is more severe in high temperatures (> 23 °C), higher relative humidity (80%) and frequent rainfall (>130 mm) (Flores, 2012).

Damage. In very severe attacks, the infection causes the fruit to crack, which causes economic losses, which can be 70% in summer; or greater than 95% during the transition of seasons from summer to winter or from winter to summer, because high relative humidity prevails. Likewise, the disease causes premature fruit drop, when they are only in the initial stage of development and the severity levels in the epidermis exceed 60% of the affected area (Insuasty et al ., 2005).

Integrated management. Weed control limits the rapid spread of the disease; I also know recommends pruning infected fruits and collect the that are thrown and bury or destroy them; Likewise, prune unproductive or dry branches, because they are a source of inoculum. Copper compounds help control the disease (Flores, 2012).

Trichoderma spp ., commercial plant extracts and chemical fungicides were used to evaluate their effects on the growth of the fungus. in conditions laboratory, for this reason, these three disease control strategies are broadly described below.

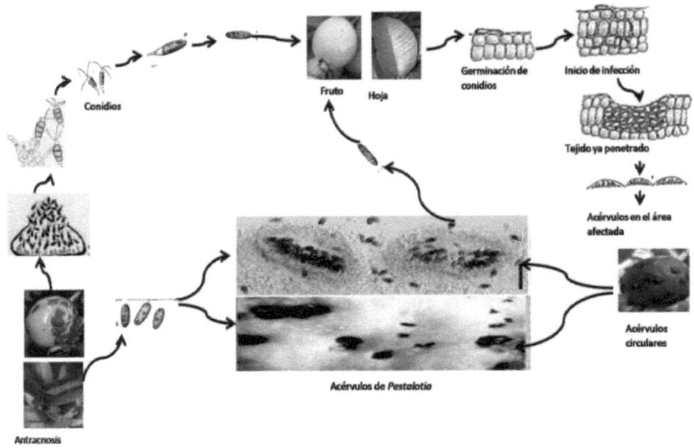

figure 7. cycle biological of Pestalothiopsis spp . (Adapted of Insuasty etc al ., 2005).

4.2. Control biological of diseases by Trichoderma spp.

Background. The Trichoderma fungus is one of the most used in the biological control of foliar and soil phytopathogens in various crops; due to its ubiquity, ease of being isolated and cultivated in numerous substrates, where it grows rapidly and does not attack higher plants (Ezziyyani, 2004). In addition, native isolates can be easily obtained from the soil where the crop grows.

Action mechanisms. It acts by competing for space and food (substrate), it also produces fungitoxic substances, induces resistance through phytoalexins and infects pathogenic fungi (Ezziyyani, 2004). Some of these mechanisms are described below.

Competence . The shortage or limitation of a requirement (space and/or nutrients), causes competition between microorganisms (Martínez et to the ., 2013). So by example HE ha It has been reported that Trichoderma is an excellent competitor against soil-dwelling fungi such as Fusarium (Hernández et al. , 2009).

Mycoparasitism . This process is complex and occurs in stages of: Chemotrophic growth , where Trichoderma can detect its possible hosts from a distance. Specific recognition of the substrate. Adhesion and coiling due to the interaction of recognition sugars on the wall of the antagonist, with lectins present on the wall of the pathogen. Lytic activity , through extracellular enzymes, mainly chitinases, glucanases and proteases, which degrade the cell walls of the fungus and promote the penetration of Trichoderma hyphae . Mycoparasitism involves coiling, penetration, vacuolization, granulation, coagulation, disintegration and lysis of the pathogen's mycelium; although all these events happen interactively and depend on the pathogen, isolation of Trichoderma and environmental conditions (Martínez et al., 2013). It has been reported he mycoparasitism exercised by Trichoderma against Fusarium , Rhizoctonia and other pathogens (Harman et al ., 2004).

Antibiosis . The metabolites with antifungal activity secreted by Trichoderma are volatile (harzianic acid) and non-volatile (viridin) compounds, with diverse structure and function, which inhibit other microorganisms, without establishing physical contact (Martínez et al.,2013).For example, Trichoderma presents antibiosis against Pythium, Fusarium, Rhizoctonia, Cylindrocladium, Thielaviopsis, Sclerotinia sclerotium,Fusarium,Helminthosporium,Armillaria,Colletotrichum, Pseudoperonospora, Verticillium, Thielaviopsis and Rosellinia (Martínez, 2012).

Advantages. Among the conveniences of using this beneficial mushroom (Anonymous, 2015b), the following stand out:

- Is efficient for he control of diseases of floors.
- Has a broad range of action.
- It has a wide capacity for colonization and reproduction in the soil, and exerts lasting control over phytopathogenic fungi.
- Help to break down subject organic, making the nutrients are converted in plant-available forms; therefore, it has an indirect effect on crop nutrition.
- It stimulates the growth of crops because it secretes metabolites that promote plant development.
- It accelerates the decomposition and maturation of compost or organic matter that contains the fungus as a biofungicide agent.

• Favors the proliferation of organisms beneficial in he floor.

• No need interval of security for Harvest .

• Preservation of the half atmosphere by decrease he use of fungicides chemicals.

• decreases the costs of production of crops.

• Attack pathogens of the root (Pythium, Fusarium, Rhizoctonia) and of the foliage (Botrytis ,

Colletotrichum , Alternaria).

• Improvement absorption of water and nutrients.

• It is not phytotoxic.

• Solubilizes nutrients in he floor for he exploitation of the plants.

• Take action as biodegradable of agrotoxics.

• Protects the seeds and material propagative of phytopathogens.

• Is compatible with mycorrhizae, Azotobacter and others biofertilizers.

• Is friendly with others microorganisms drivers of pests and diseases.

Disadvantages. Some drawbacks cited by Guedez et to the (2008), describe next .

• Ignorance about the beginning of this method of control biological.

• Lack of support economic for the acquisition of the product.

• No this easily available.

• HE have issues with economic thresholds .

• No act equal against all the fungus phytopathogens.

Taxonomy: He fungus biocontroller presents the following taxonomy :

Anamorph . The asexual phase corresponds to Trichoderma that belongs to the Kingdom: Fungi , Division: Deuteromycota , Subdivision: Deuteromycotina, Class : Deuteromycetes, Order: Moniliales and Family: Monileaceae (Romero, 1988; Michel, 2001).

Teleomorph . Corresponds to Hypocrea, which is located in the Kingdom: Fungi, Division: Eumycota, Class: Ascomycetes, Order: Hypocreales and Family: Hypocraceae (Villegas, 2014).

Morphology: The characteristics of the sexual and asexual reproduction phases are presented below. Anamorph (Trichoderma) . Of agreement with Gams and Bissett (1998), the mycelium develops phialides whorled, in alternating pairs or irregularly arranged with the terminal parts elongated and narrow. Hyphae can be wide and straight or relatively narrow and flexible. The conidia are globose, ellipsoid, cylindrical, smooth wall; with green, gray, brown, dark or pale colors. Chlamydospores are common, with globose or elliptical shapes, and terminal or intercalated positions, yellowish or greenish in color and 1-15 μm in diameter in most species (Figure 8).

Teleomorph (Hypocrea) . The sexual stage is less common in nature; It is characterized by having perithecial ascomas immersed in stromas with high morphological variability, with green or hyaline ascospores formed in typically cylindrical asci with a blunt apex, sometimes slightly thickened, with 8 mono- or dimorphic bicellular ascospores, which disarticulate at maturity, leaving 16 spores. by asca (Kullning-Gradinger et al., 2002). Chaverri and Samuels (2003) verified that this group is not monophyletic; he was not assigned e taxonomic category value.

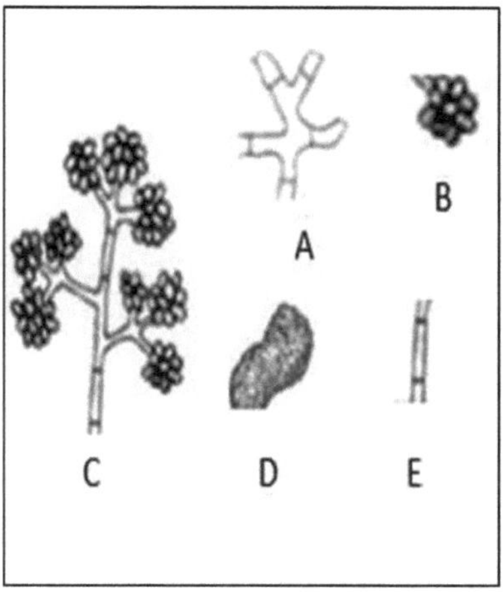

Trichoderma

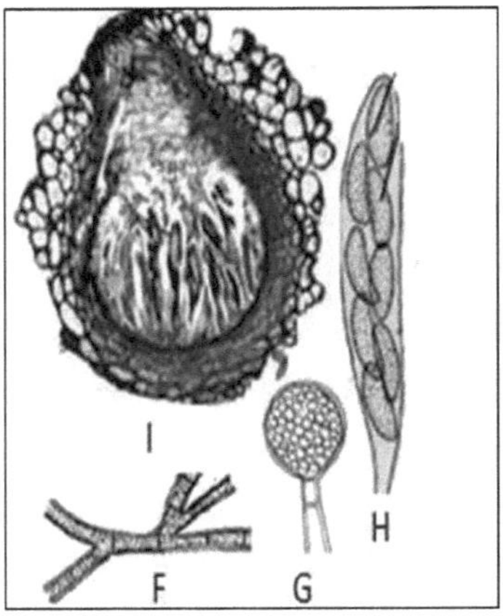

Hypocrea

Figure 8. Morphology of Trichoderma (left): A) Mycelium, B) Conidia, C) Conidiophores, D) Sclerotia, E) Phialide, and Hypocrea (right): F) Hyphae, G) Chlamydospora, H) Asca with ascospores , I) Perithecium with asci (Anonymous, 2014c).

Cycle biological. The chlamydospores are the structures of survival of Trichoderma spp . in adverse conditions; It lives as a saprophyte in soil and wood. Moisture in the soil (60%) causes the fungus to go from a dormant to an active state; but at higher levels of saturation, colonization and survival decrease due to the low availability of oxygen; Likewise, at acidic pH the reproduction of the fungus increases (Figure 9) and it colonizes the roots of plants to live in symbiosis (Villegas, 2014).

Commercial products made with Trichoderma . Some commercial biological products made with Trichoderma spp. are presented in Table 1 .

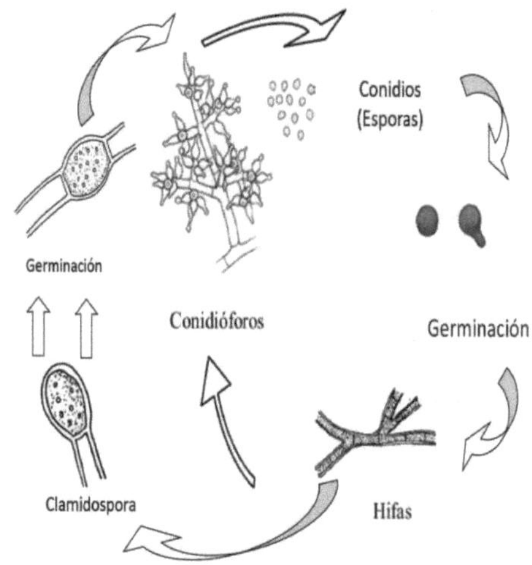

Figure 9. Cycle biological of Trichoderma (Villegas, 2014; Plancarte, 2014).

Chart 1. Products biological commercial that they contain Trichoderma spp.

Active ingredient	Tradename	Pathogen that controls
Trichoderma harzianum Rifai	BIOBEN	Phytophthora capsici Leonian.
	NATU CONTROL	Phytophthora,Rhizoctonia,Pythium,Sclerotinia,Sclerotium,FusariumyVerticillium
	PHC T-22 ®	Pythium, Rhizoctonia, Fusarium, Cylindrocladium,Thielaviopsis, Verticillium y Sclerotinia.
	TRICHO-SIN WP	Damping off, Fusarium
	TRICON	Sclerotinia, Helminthosporium, Armillaria, Colletotrichum, Pythium, Phytophthora, fusarium Rhizoctonia y Pseudoperonospora.
Trichoderma spp .	spectrum I knit BIO	Rhizoctonia, Fusarium, Pythium, Botrytis, Phytophthora, Sclerotium, Sclerotinia, Alternaria Colletotrichum Thielaviopsis, Armillaria y Roselinia.
Trichoderma virens (JHMill., Gidde ns&A.A. Foster)Arx	PHC ROOTMATE	Phytophthora.

Fountain: Thomson (2012)

Description of products with Trichoderma used in research. In this study, the seven products described below were selected :

Bactiva MR . It is a dry and soluble biofungicide that prevents root rot due to pathogenic fungi. The product contains Bacillus polymyxa Prazmowski highly antagonistic to Fusarium; also B. subtilis Cohn and several species of the beneficial fungus Trichoderma ; These compete with pathogens for space and nutrients in the rhizosphere; They also parasitize them and prevent their growth by secreting toxins. It's a biorooter; that is, a stimulant of the growth of the root system because it contains beneficial bacteria and fungi that produce growth hormones (gibberellins, cytokinins), such as B. megaterium Bary and Pseudomonas flourescens Migula; which, together with Trichoderma stimulate the growth of the root system throughout the entire growth cycle. crop and are essential during the emergence of seedlings or in the recovery of damaged roots. It is a microbial biofertilizer due to its nitrogen-fixing bacteria and phosphorus solubilizers; that fix or release elements present in the soil, but that the floors No they can leverage. The microorganisms beneficial are supplemented by biostimulants that promote their biological activity, increase the propagation of cells and Lateral buds also delay the aging of plant tissues. The product It is composed of beneficial nitrogen-fixing bacteria, which solubilize phosphorus and produce plant growth hormones and vitamins (biotin, folic acid, B, B2, B3, B6, B7, B12, C, K); also contains amino acids, vegetable protein, soluble yucca extracts (Yuccaschidigera L.) and seaweed (Ascophyllum nodosum L.); fulvic acids and derivatives of leonardite (DEAQ, 2015).

Biobravo ® . It is an alternative to the use of chemical fertilizers; contains microorganisms growth promoters called biofertilizers, such as bacteria of the genus Azospirillum , the mycorrhizal fungi of the genus Glomus, and the biocontroller Trichoderma sp. Fungi are important in plants because they are associated with roots, form a bidirectional transfer system, carrying mineral nutrients and soil organic compounds. Azospirillum has the ability to fix nitrogen from the air, for the use of the plant (DEAQ, 2015).

PHC T-22 ® . It is a preventive biological fungicide for the control of diseases of numerous plant species. The active ingredient is a beneficial microorganism, T. harzianum Strain T-22 (KRL-AG2), which can be applied to the soil using a drip system (DEAQ, 2015).

FITHAN. Its active ingredient is T. fasciculatum with a concentration of $1.3x10^{6}$ spores per liter; It is a moisturizing solution, this product controls root rot, wilting, drowning, canker and fungosis development; Its mechanism of action is mycoparasitism. It is applied in the drip irrigation system or by sprinkling with a pump (Carrasco, 2014) .

T. asperellum (Chilapa strain) . This strain was provided by the laboratory of CEP-CSAEGro Phytopathology. He isolated himself originally from a soil sample extracted from a field cultivated with pipiana squash (Cucurbita argyrosperma Huber.), in the town from El Refugio, municipality of Chilapa, Gro., in order to test the antagonistic effect (antibiosis) on the growth of phytopathogenic fungi that rot the fruit of this cucurbitaceae.

T. asperellum (Cocula strain) . This strain was provided, like the previous one, by the Phytopathology laboratory. It was originally isolated from a soil sample obtained in the agricultural field of the CEP-CSAEGro. It has been tested in various investigations on control biology of pathogenic fungi. For example, Díaz (2013) obtained an average of 100% of inhibition of P. capsici Leonian.

Effect of Trichoderma on Pestalotia. It has been observed that the application of Trichoderma at a dose of 80 g 100 L^{-1} of water, with 60 mL of adherent and dispersant, efficiently controls the disease caused by clove P. microspora (Flores, 2015).

On the other hand, Ortiz (2015) compared the effect of different strains of Trichoderma in laboratory conditions, finding favorable results when 100% inhibited the strains of T.asperellum (chilapa strain), Trichoderma spp and T. reesei, since they exerted greater competition for space and nutrients and interacted directly with the pathogen (mycoparasitism).

Likewise, Rayo (2014) carried out an investigation with 5 treatments and 4 repetitions comparing different strains of trichoderma, obtaining favorable results since the strains of Bravo 1 and Bravo 2 are the ones that managed to inhibit 100% and 50% of the growth of the colony. fungi as they are good for nutrient competition.

4.3. fungal diseases

Background. The use of natural products is as old as humanity; They are of a very varied nature, from biologically active substances of marine origin, root exudates and secondary metabolites obtained from plants, of which the properties and active ingredients have mainly been studied, because their use is an alternative to produce safe foods, and helps balance between the environment, production and the man. HE they tried with success in The struggle against diseases and pests, extracts of neem (Azadirachta indica L.), garlic (Allium sativum L.) and others (Rosales et al ., 2002).

Importance. They are a promising option to be used as fungicides or regulators of fungal development. phytopathogens; in it frame of the agriculture sustainable, because The compounds obtained from plants are biodegradable, highly effective, low cost and do not contaminate the environment (Martínez et al ., 2010).

mechanisms action. Their properties antimicrobial HE they must to the effect of secondary metabolites such as alkaloids, essential oils, polyphenols, tannins, saponins, anthraquinones, flavonoids, terpenes, lectins, polypeptides and phenols. For example, the toxicity of phenols is attributed to enzymatic inhibition by oxidation of compounds in microorganisms; Likewise, terpenes and essential oils cause the breakdown of the membrane through lipophilic compounds; alkaloids intercalate with DNA; lectins and polypeptides can form ion channels in the membrane (González et al., 2008).

Types of commercial products. Some examples of organic products available on the market are described in Table 2.

Chart 2. Products commercial elaborated with extracts vegetables, for he control of phytopathogenic fungi

No.	Product	Ingredient asset	Patagonia que control
1	AGRILIFE	Acid ascorbic	Pythium de baryanum R. Hesse . y translations botrytis cinerea (De Bary) Whetzel.
2	PROGRANI C	Extracted about cinnamon (innamon zeylanic L.)	R. solani Kühn , Fusarium spp. Pythium spp .,Oidium spp. Erysiphe spp .
3	PROGRANI C ®	Extracted about take (A. indicate Juss)	R. solani Fusarium spp. Pythium spp.
4	SEDRIC 4 times	Yucca (manihot food L.)	solani , Erwinia carotovora (Jones.) Waldee , Alternaria solani Sorauer , P. infestans De Bary , Xanthomonas campestrisp v. vesicatoria, P. capsici Leonian , Verticillum sp., Sclerotinia sclerotiorum (Lib.) de Bary , Fusarium spp., Colletotrichum spp., Alternaria spp . y Erwinia amylovora (Burrill) Winslow and to the.
5	CANOL ®	Extract of cinnamon (C. zeylanicum L.)	Oidium spp .
6	TAIL OF HORSE ®	Extract of line of horse (Equisetum arvanse L.)	Phythophthora spp ., Oidium spp., Botrytis spp., Alternaria spp.y Pestalotia psidii Pat.
7	BIO 150 CITRUS ®	Extract of seeds of citrus	Ascochyta spp., Fusarium spp., Botrytis spp., Alternaria spp., Rhizoctonia spp ., Sphaeroteca spp ., Mycosphaerella spp . y Colletotrichum spp.
8	BIO 75 TOMOL®	Extract of red thyme (Thymus vulgaris L.)	Ascochytaspp., Fusarium spp., Botrytisspp., Alternariaspp., Rhizoctoniaspp., Sphaerotecaspp., Mycosphaerellaspp., Colletotrichumspp., Cercosporaspp.,

			Septoriaspp., Stemphyliumspp., Peronosporasp. y Pythiumspp.
9	LECITIN ®	Lecithin de soya (Glycine max L.)	Puccina spp .
10	LIPPOIL ®	Extract de Lippia graveolens Kunth.	Erwinia sp., Downy mildew sp ., Alternaria sp ., Cercospora sp ., Colletotrichum sp ., Aspergilliu ssp .y Penicillium sp.
11	CINNOIL ®	Extract de Canela (C. zeylanicum L.)	Erwinia sp ., Downy mildew sp ., Alternaria sp ., Cercospora sp ., Colletotrichum sp. Aspergillius sp.y Penicillium sp.
12	CAPSIOIL®	Extracted from Ajo (A. sativum L.) , oil about Chile (Capsicum annual L.) and of cinnamon (C. zeylanicum L.)	Erwinia sp ., Peronospora sp . and Alternaria sp.

Fountain: Thomson (2012).

Description of the products used in the investigation. In he present study Six commercial products described below were used.

CINNOIL. Product that can be used in integrated pest and disease management, as well as in conventional agriculture due to its completely natural origin (Table 3). When sprayed on foliage, it penetrates through the stomata until the fabrics of driving, protecting completely the plant (Thomson, 2012).

Chart 3. Composition chemistry of CINNOIL

Composition	% in weight
Oil of cinnamon (C. Cassia and c. zeeylanicum)	
Aldehyde cinnamic and Eugenol No less of	50.0
Oil of L. graveolens k L. and L. berlandieri I ..	15.0
Extract essential of garlic Allium spp. No less of	10.0
Adjuvants organic No less of	10.0
Diluents natural No further of	10.0
Total	100.0

Fountain: Thomson(2012).

LIPPOIL. This product can be used in he driving integrated of pests and diseases, So like in the Agriculture conventional due to his origin completely natural (Table 4). It is a bactericidal and fungicide that, when applied foliarly, inhibits the germination of spores and the development of phytopathogenic fungi and bacteria. It is used to control diseases caused by bacteria of the genera Xanthomonas sp ., Pseudomona sp. and Erwinia sp **.,** (Thomson, 2012).

Chart 4. Composition chemistry of LIPPOIL

Composition	% in weight
Oil of L. graveolens L. and L.berlandiere L, no less of	40.0
Oil of cinnamon (C.cassia and zeylanicum) no less of	40.0
Adjuvants organic No further of	20.0
Total	100.0

Fountain: Thomson (2012).

CAPSIOIL. Is a repellent and insecticide natural Elaborated mostly to leave of active ingredients of high concentration and purity (Table 5);It is very effective for the integrated management of a broad cluster of pests. Can be used only either in pesticide mix biological either chemicals, to improve the effectiveness of applications; as well as in organic or ecological agriculture (Thomson, 2012).

Chart 5. Composition chemistry of CAPSIOIL

Composition	% in weight
Oil of chili c. frutescens L. and c. annuum L. No less of.......	40.00%
Extract essential of garlic Allium spp . No less of	30.00%
Cinnamon oil (C. cassia and C. zeeylanicum) No less of	20.00%
Adjuvants organic No less of.....	10.00%
Total	100.00%

Fountain: Thomson (2012).

Chart 6. Composition chemistry of NEEM CINNOIL

Composition	% in weight
Oil of neem (TO. indicates) Azaridactin not less of	7.5
Oil of cinnamon (C. Cassia and c.zeeylanicum) Aldehydocinnamic and Eugenol No less of less of	40.0
Adjuvants organic No less of......	30.0
Diluents Natural No further of	22.5
Total	100.0

Fountain: Thomson (2012).

NEEM CINNOIL. It is a botanical pesticide obtained from the A. indica plant (Table 6). It can be sprayed on crops as an organic substitute, instead of chemical insecticides; It is effective for the prevention and control of several fungal diseases such as powdery mildew, downy mildew, black spot, Botrytis , anthracnose, rust, leaf spot, phylloxera and Alternaria (Thomson, 2012).

BIOCINNAMON. Acts as a fungicide, bactericide and antiparasitic (Chart 7). Stimulates the systems

Chart 7. Composition chemistry of BIOCINNAMON

Composition	% in weight
Cinnamon (C. zeylanicum).	30.0
Pepper black	20.0
Solvents, emulsions, adjuvants and compounds related	50.0
Total	100.0

Fountain: Thomson (2012).

Chart 8. Composition chemistry of ALLIOIL

Composition	% in weight
Extract essential of Garlic Allium spp. No less than	90.0
Diluents and conditioners organic No further of	10.0
Total	100.0

Fountain: Thomson (2012).

Natural defense of the plant and vegetative regeneration, against climatic or biotic stress. It is effective in preventive treatments against powdery mildew and mites, in doses of 200-250 mL L- 1 , in. Foliar sprays done in the afternoon with little solar radiation. It should not be mixed with products of alkaline or acid reactions (Thomson, 2012).

ALLIOIL. It inhibits the growth of phytopathogenic fungi (Table 8); it has been proven to combat fungi : Penicillium italicum Wehmer , Aspergilus flavus Link. , Fusarium sp., R. solani , Alternaria sp., Colletotrichum sp ., Pythium sp. and others (Thomson, 2012).

Effect of plant extracts used against Pestalotiopsis and other fungi. There is little information about the use of these organic products in field crops (in situ); For example, it was reported that ethanolic crude extract obtained from young eucalyptus leaves (Eucaliptus globulus L.), in biweekly sprays applied to the crown of guava trees, in doses of 8 mL L^{-1} of water, managed to reduce the incidence of P. microspora by 94.7% (Insuasty et al., 2005).

However, under laboratory conditions (in vitro), extracts of eucalyptus , mint (Menta piperita L.) and horsetail (Equisetum arvense L.) have been tested on the development of P. microspora colonies ; It was found that the eucalyptus product showed the best control, inhibiting 80% of the growth of the fungal colony (Quijada and Gómez, 2005).

Likewise, it was found that the neem extract inhibited 100% the growth of P. microspora , and maintained its residuality for 9 days in PDA culture medium (Rayo, 2014); while in another similar trial, it was reported that the Reynoutria sachalinensis (F. Schmidt) Nakai extract completely inhibited (100%) the growth of the colonies of this pathogen grown in PDA (Ortíz, 2015). In other experiments carried out with species of Colletotrichum , which is a pathogen of parts aerial of the plant, like Pestalotiopsis , is has reported that Plancarte (2014), tested different extracts under the same conditions, obtaining favorable results with the lippia and cinnamon extract.

4.4. Use of fungicides chemicals for check diseases

Background. The technology of the green revolution promoted the use of improved varieties, the application of large amounts of water, fertilizers and pesticides to increase agricultural productivity. Since then, synthetic fungicides are the most used because they present the advantages of: increasing yields, they are fast acting; They have affordable prices; They are found in multiple presentations; They are easy to get in the markets; They have a wide range of action and there are many active ingredients (Avalos, 2014).

Advantages . Synthetic fungicides are the most used products because they have the advantages of: increasing yields, they are fast acting; they have prices accessible; They are found in multiple presentations; They are easy to get on the market; They have a wide range of action and there are many active ingredients (Avalos 2014).

Disadvantages . He use irrational of this method of control of diseases, ha caused environmental contamination, residual effect, toxicity, pH imbalance in the soil, resistance in fungi and health problems in humans and animals; Therefore, its usefulness is currently negatively questioned (Avalos 2014).

Mechanisms of action in the cell . The fungicide has a direct effect on the biology of the microorganism; acts on the structure of the cell membrane of fungi causing the loss of cellular content and being responsible for death of the fungus (Ayala, 2008). For example, copper interrupts the transport of energy.

Classification of fungicides. Fungicides are classified as follows, Table 9.

Recommended fungicides against Pestalotiopsis. There is no wide range of chemicals recommended for controlling Pestalotiopsis species; however, it recommends applying copper-based products, Kasugamicin, Zineb and Mancozeb (Insuasty et al., 2005). However, it is considered that the fungicides recommended against Colletotrichum also control the first fungal genus; Some of them are described in Table A-1.

Chart 9 . Classification of the fungicides by he mode and the shape of action.

Action mode	Classification	Form of action
Distribution (mobility in plant)	Protectants (contact)	They remain on the surface of the plant where they are applied. They do not have any control activity after infection. They require frequent applications every 7 to 14 days to protect new areas of plant growth.
	Systemic	Plants absorb the fungicide through the foliage or roots and generally translocate it upwards through the xylem and rarely downwards through the phloem. They offer control after infection
	Translaminar	They penetrate the sheet from he beam toward he Instead, they are translocated locally
In he cycle of mushroom life	Preventive	It is present in the plant as a protective barrier before the pathogen arrives and develops.
	Healing	Stops the development of the pathogen in the plant tissues; They are effective 24 to 72 hours after infection.
	Antisporulants	Prevents the development of spores. The disease continues to develop but the product prevents the release of the spores.
Spectrum of activity	Unisite	Activity fungal against a only spot in the metabolism of the fungus.
	Multisite	It affects different sites metabolic inside the fungus.

Fountain: Ayala (2008)

Description of fungicides used in this research. In the present study, only six fungicides were selected to evaluate them in vitro against the fungus P. microspora , which causes anthracnose in guava; which are described below.

Chart 10. Composition chemistry of BENOMYL

Composición	% en peso
Ingrediente activo:	
benomil: metil-1 (butilcarbamoil) bencimidazol - 2- ilcarbamato. No menos de (Equivalente a 500 g de i.a. kg-1)	50.0
Ingredientes inertes : Dispersantes, humectantes, secante y diluyente. No más de	50.0
Total	100.0
Fuente: Thomson (2012)	

BENOMYL (benomyl). It is a fungicide with systemic and contact action; It is residual; It is applied as a spray for the prevention and control of various fungal diseases (Table 10) and is compatible with most commercial insecticides and fungicides (Thomson, 2012).

CERCOBIN (thiophanatomethyl). Systemic thiocarbamate with fungicidal, preventive and curative activity, systemically and by contact on diseases caused by fungus endo and ectoparasites (Chart eleven). Its spectrum of activity is extraordinarily broad and includes, among others,diseases produced by

Chart 12. Composition chemistry of MANZATE

Composition	% in weight
Ingredient asset:	
mancozeb: (product of coordination of the zinc ion and ethylenebisdithiocarbamate of manganese).No less of........................	80.0
(Equivalent to 800 g of ia kg- 1)	
Ingredients inert :	
adjuvants, humectants. No further of ..	20.0
Total	100.0
Fountain: Thomson (2012).	

Chart eleven. Composition chemistry of CERCOBIN

Composition	% in weight
Ingredient asset:	
Thiophanate methyl: Dimethyl -4, 4- 0-phenylenebis (3-thioalophanate). No less of ...	70.0
(Equivalent to 700 g of ia kg - 1)	
Ingredients inert :	
humectants, dispersants, impurities and compounds related. No further of	30.0
Total	100.0

Fountain: Thomson (2012).

Botrytis type fungi , anthracnose, cladosporiosis, screening, Fusarium , Monilia, mottle, powdery mildew, rusts and septoria (Thomson, 2012).

MANZATE (mancozeb) .Is a fungicide of contact and preventive; comes of the combination of the zinc ion with maneb (Table 12). It is used in doses of 0.5 to 1 kg dissolved in 200 to 100 and 50 to 60 L of water ha $^{-1}$, to be applied with ground and aerial equipment, respectively. The exclusive use of any fungicide can lead to the development of resistant races of fungi that manifest as a decrease in the level of disease control (Thomson, 2012).

CUPRAVIT (copper oxychloride). It is a wettable powder (Table 13); act against various diseases; It is incompatible with several products alkaline. It should be applied preventively, before the first symptoms appear or are detected; can be used without fear of inducing endurance in pathogens and allows the harvest to be exported (Thomson, 2012).

ZINEB (zinc ethylene bis-dithiocarbamate).a fungicidedithiocarbamic

Chart 13. Composition chemistry of CUPRAVIT

Composition	% in weight
Ingredient asset:	
Oxychloride of copper . No less of.........	85.0
(Equivalent to 850 g of ia kg- 1)	
Ingredients inert :	
Thinners, moisturizers and dispersants.No further of.	15.0
Total	100.0

Fountain: Thomson (2012).

Chart 14. Composition chemistry of ZINEB

Composition	% in weight
Ingredient asset:	
Ethylene Bis (dithiocarbamate of zinc). Noless of	80.0
(Equivalent to 800 g of ia kg - 1)	
Ingredients inert :	
Thinners, moisturizers and dispersants.No further of..	20.0
Total	100.0

Fountain: Thomson (2012).

Chart fifteen. Composition chemistry of CAPTAN

Composition	% in weight
Ingredient asset:	
They capture: N- (trichloromethyl) uncle-4 cyclohexen - 1, 2 dicarboximide. No less of..	50.0
(Equivalent to 500 g of ia kg - 1)	
Ingredients inert :	
diluent, moisturizers, dispersant and compounds related. No further of	50.0
Total	100.0

Fountain: Thomson (2012).

Formulated as a wettable powder, which dissolves in water to be applied as a direct spray (Table 14). This product acts through contact in the prevention and control of diseases (Thomson, 2012).

CAPTAN (captain). It is a contact fungicide. Its chemical composition is shown in the Chart fifteen. No is compatible with THIRAM (thiram), polysulfides, compounds mercurials, DIAZINON (diazinon), lime and agricultural oils; It should not be applied during hours of intense heat or when wind speed is high. (Thomson, 2012).

Effect of chemical fungicides against Pestalotiopsis . This disease can be prevented with sprinkles foliar of products to base of copper in dose of 300 to 400 g 100 L $^{-1}$ of water (López et al., 1994).

In studies carried out in the Hoya de Río Suárez to define control strategies for the disease, it was found that the use of Kasugamicin (kasumin), a fungicide product, carried out good control because it reduced the progression of the disease in affected fruits, the incidence and protection of healthy fruits. The effective application frequency is with a dose of 3 mL L^{-1} of water, up to 30 days before harvesting. In general, for sprinkle a tree of height average of 3 m, HE requires spraying 1.5 L of fungicide solution (Insuasty et al., 2005).

Lopez (1994) mentions that in the Calvillo area, Aguascalientes, clove disease can be prevented with the foliar application of copper-based products, such as Cupravit® pH 85% (copper oxychloride) in doses of 300-400 g 100 L^{-1} of water or Zineb® pH 80% (Zinc ethylene bis dithiocarbamate) 80 pH, using 120-180 g 100 L^{-1} of water.

Prakash and Pandey (2007) point out that in India, clove control is achieved with three or four applications of 0.3% copper oxychloride, with intervals of 15 days. In Costa Rica, the fight against this disease is carried out with products derived from copper, carbamates and benomyl, after flowering.

In " in vitro" tests, it was observed that 50% copper oxychloride at doses of 100g L^{-1} of water No reduced he development from Pestalotiopsis microspore , while that he Chlorothalonil 75% at doses of 50 and 100 g L^{-1} of water, had the best result, based on these results, it was concluded that Chlorothalonil, or the combination of mancozeb and metalaxyl, mancozeb or carbendazim, can be used to control the disease (Ray et al., 2007). Kasar et al., (2006) evaluated "in vitro" fungicidal products against Pestalotiopsis microspora ; They concluded that the carbendazim + thiram mixture was the most effective in inhibiting mycelial development, followed by chlorothalonil and mancozeb; while carbendazim, metalaxyl + mancozeby and copper oxychloride were not effective against the pathogen.

V. MATERIALS AND METHODS

5.1 Characteristics of the area of study.

The present research was carried out in the Phytopathology Laboratory of the Center for Professional Studies of the College Superior Agricultural of the State of Guerrero (CEP-CSAEGro); located at kilometer 14.5 of the Iguala-Cocula highway, between the coordinates 18° 14' 00" north latitude and 99° 40' 00" west longitude of the meridian of Greenwich; at 640 m altitude (Figure 10). In this region predominates a climate dry tropical Awo (w) (i) g , with rains in summer. Average precipitation and temperature annual are of 797 mm and 26.4 °C, respectively (García, 1973).

5.2 Sampling in field.

Samples of infected fruits of guava plants (Figure 11), in a local orchard from Tetipac, Gro. The sick material was saved in bags of paper and HE transferred to the Phytopathology laboratory.

5.3 Isolation and purification of the fungus.

Figure 10. Location of the place experimental.

Figure eleven. Fruits infected in plant of guava

The humid chamber technique was used to induce sporulation and/or mycelium growth, in pieces of diseased tissues, using the following procedure: Infected fruits HE they washed with water and soap, HE they cut pieces of 0.5 mm 2 including parts of diseased and healthy tissue were disinfected with 1 % sodium hypochlorite for 1 minute, rinsed three times in water distilled sterilized, HE they placed five pieces about two slide previously flamed, ready in shape of cross about paper absorbent moistened with water distilled inside of a box Petri; after, this It was sealed and labeled with data on: name, date and type of fabric; It was incubated for 4 to 5 days at room temperature (+ 28 °C) in the laboratory (Figure 12).

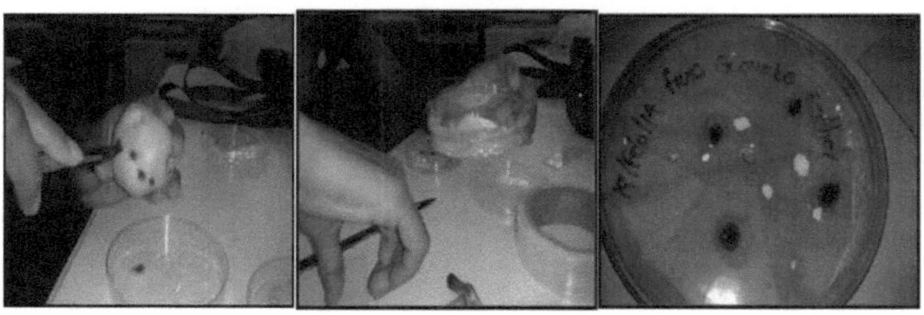

Figure 12. Preparation of camera wet with tissue infected, of fruit of guava, in box Petri.

The pieces of infected tissue placed in the humid chambers were reviewed under the stereoscopic microscope, the mycelium and acervuli of the fungus were located; samples were taken of conidia and HE prepared a suspension of these with water sterilized on an excavated slide; A sample of the suspension was transferred with a bacteriological loop and streaked on the PDA surface; It was incubated for 3 to 4 days, the culture was inspected, small monosporic colonies were located; one was selected, tips were cut of hyphae, HE transferred to P.D.A. and HE got he crop monosporic of the fungus (Tuite, 1969); that HE preserved in a tube of rehearsal with P.D.A. for use it in subsequent trials (Figure 7).

5.4. ID morphological of the fungus .

Of the monosporic strain HE they took samples of mycelium and conidia, it became a preparation in lactophenol, was placed in the compound microscope and the characteristics were observed of: shape, color and septation of are structures (Figure 13), So as the type of fruiting bodies produced in the culture medium; which served as the basis for identify to the fungus, taking as reference the keys illustrated of Barnett and Hunter (2000).

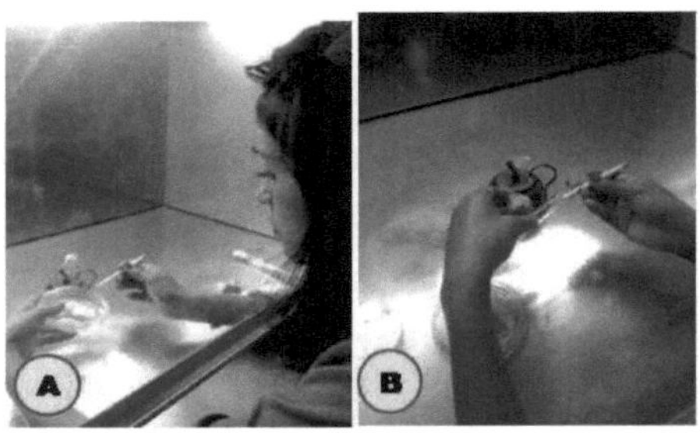

Figure 13. TO) Sowing of the fungus in P.D.A. and B) Transfer of the fungus purified to P.D.A. in tube of rehearsal.

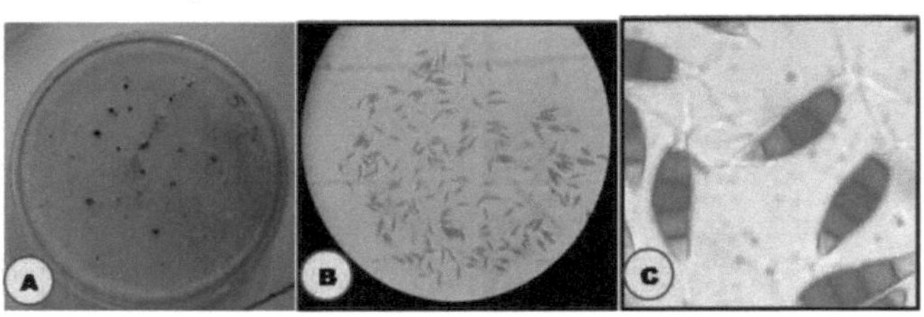

Figure 14. TO) Fungus purific9786139404391ado acervuli black (TO), conidia of the fungus (b) (B: 40x; C:100x)

5.5. Molecular identification of the fungus. From the monosporic culture, the extraction was carried out of acid deoxyribonucleic mycelial, using a sample of fifty to 100 mg of mycelium; the DNeas and MR kit was used (Qiagen, 2012); 4 repetitions were carried out. They were carried out reactions in chain of polymerase (PCR) universal for fungus with the oligos ITS-1fu 5'-tccgtaggtgaacctgcgg-3' and ITS-4 5'-5tcctccgcttattgatatgc-3' according to White et al ., (1990); which amplify an internal intergenic spacer (ITS) generating a product of variable size, approximately between 500 and 900 base pairs (bp). This practice HE performed in a volume of 25 µL of a mix with shock absorber 1X reaction , MgCl 2 2 mM, dNTP's 200 nM of each, 20 pmol of each oligonucleotide and 1 unit by Taq DNA polymerase. The program thermal consisted in maintaining a temperature to 94°C during 2 min, Following of 35 cycles to 94-55-72 °C during 30-30-60 sec and a final extension of 5 min at 72 °C. The products of the PCR reactions were separated by electrophoresis in gels of agarose to 1.5 %, and the bands HE they observed in a UVP brand ultraviolet light transilluminator and photographed. The amplified fragments by PCR were sequenced directly and HE they photographed. They were compared the results with the sequences available in he bank of genes (Gen Bank) of the National Center for Biotechnology Information (NCBI) of the National Institute of Health (NIH) of the USA through a BLAS (Díaz et al., 2015).

5.6 Proof of pathogenicity.

To verify the pathogenicity of the isolated fungus, Koch's postulates were carried out, using the following material and inoculation techniques:

Crop of fungus. The cologne of the fungus cultivated in P.D.A. had a age approximate of 1 month, in where HE they observed acervuli with dark masses of conidia.

Healthy fruits. These were obtained from a disease-free orchard. They were taken to the laboratory, first the tray was disinfested with the help of cotton with alcohol where would placed; in the bottom HE placed a layer of cotton (Figure 13), which was moistened by sprinkling it with distilled water to maintain favorable humidity for fungal infection. The healthy fruits were washed with soap and water, cleaned with alcohol (96%) and slightly exposed to the flame of the lighter to ensure complete disinfection of the epicarp.

Healthy leaves. They were collected in the same orchard from which the fruits were obtained, they were chosen the that were further healthy without damage either symptoms of the disease, they took to the laboratory and HE performed he same procedure of cleaning that in the fruits.

A. Techniques of inoculation in fruits. HE performed as HE indicates to continuation

Treatment 1: Punch wounds. A 0.7 mm diameter punch was used. diameter, HE pressed and turn gently for cut and withdraw part of the epicarp in 4 sides (points cardinals); in each orifice HE inoculated with a disk (0.7 mm of diameter) of PDA+fungus (Figure 15). 4 repetitions were done.

Control treatments 1: punch wounds. The wounds on the 4 sides of the fruit were made with punches; PDA discs (culture medium only) of equal diameter were cut and placed covering the wounds in the 4 sites (Figure 15). 4 repetitions were made.

Treatment 2: Wounds with needle of dissection. HE they made with needle flamed, wounds in 4 places (cardinal points) of the fruit; A suspension of mycelium and conidia was prepared in sterile distilled water; For this, a slide was used to scrape the surface of PDA and remove the fungal colony, it was transferred to a beaker, and the volume was measured. to fifty mL, HE took 1 mL, HE placed a sample in the camera of Neubauer and HE determined the concentration of 6.8 $x 10^6$ conidia by mL; by last, HE sprinkled with a manual sprayer (1 L) the inoculum of the fungus, completely covering the epicarp of the fruit (Figure 15). 4 repetitions were done.

Control treatments 2: Wounds with dissection needle. Wounds were made in 4 places on the fruit (cardinal points) and sterile distilled water was sprayed with a manual spray bottle, covering the entire fruit (Figure 15). 4 repetitions were made.

Treatment 3: Spraying conidia suspension. The inoculum suspension contained in the atomizer was sprayed completely covering the healthy fruits (without wounds) (Figure 15). 4 repetitions were done.

Control treatment 3: Spraying of sterile distilled water. With a manual sprayer, only sterile distilled water was sprayed. 4 repetitions were done.

Treatment 4: Inoculation with PDA+ fungus. HE they cut with he punch, discs 0.7 cm of diameter, of the cologne of fungus cultivated in PDA; HE inoculated depositing the inoculum on the surface of the fruit in 4 cardinal points (Figure 1). 4

repetitions were done .

Treatment witness 4: Inoculation with PDA. HE they cut with he punch, discs of 0.7 cm of diameter, of half of crop P.D.A. (without fungus); deposited about the surface of the fruit in 4 points cardinal (Figure 1).It they made 4 repetitions (Figure 15). All the fruits were placed on the cotton layer inside the tray; this was covered with plastic transparent by 72 hours for favor the infection of the fungus.

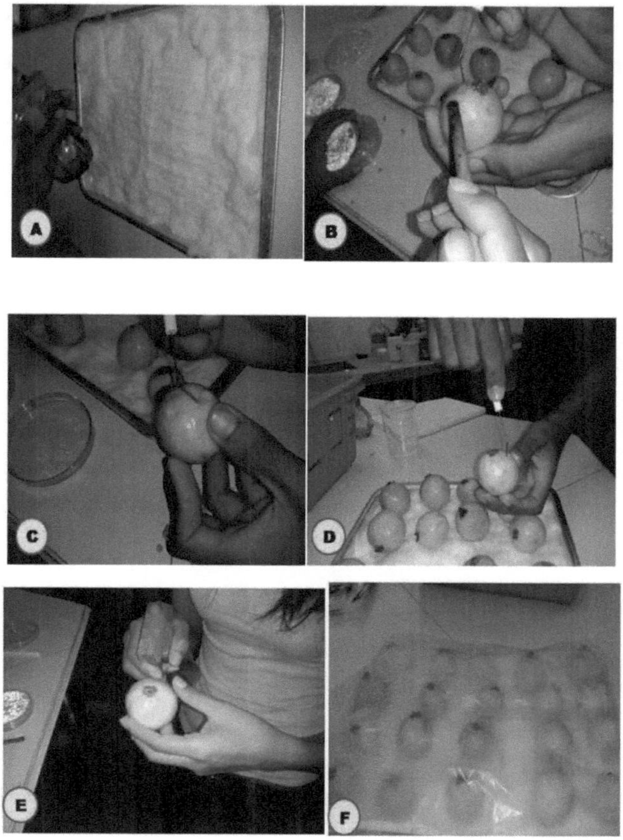

Figure 15. A) Placing the cotton in the tray. B) Placement of the PDA in the fruit C) Fungus in fruit with wounds of punch D) Fungus in he fruit with wounds of the needle AND) Aspersion of the inoculum suspension YF) Tray with fruits covered with plastic.

B. Techniques of inoculation in leaves. HE performed he procedure that HE describe next :

Treatment 1: Wounds with punched out With he punch of 0.7 mm of diameter, pressed and turn gently for cut and withdraw he tissue, in three sites of the beam of the sheet; where pieces of the same size, of PDA+ fungus, were inoculated, which were fixed with adhesive tape (Figure 16). 4 repetitions were done.

Treatment witness 1: Wounds with punched out With he punch HE short he woven in three sites of the beam of the sheet; in where HE they placed pieces of the same size, of PDA alone (without fungus), which were fixed with adhesive tape (Figure 16). 4 repetitions were done .

Treatment 2: Wounds with needle of dissection. HE they made with needle flamed, wounds in he beam of the sheet; and about are HE they placed pieces of PDA+fungus, which were fixed with adhesive tape (Figure 16). There were 4
Control Treatment2: Wounds with a dissecting needle. They were made with a flamed needle, wounds on the upper part of the leaf; and pieces of only PDA without fungus were placed on them (Figure 16). 4 repetitions were done .

Treatment 3: Inoculation with conidial spraying. The inoculum suspension was sprayed with atomizer about he limbo of leaves healthy without wounds (Figure 16). They did 4 repetitions.

Control treatment 3: Spraying of sterile distilled water. This was sprayed with a spray bottle about he limbo of leaves healthy without wounds (Figure 16). they made 4 repetitions.

Treatment 4: Inoculation with PDA + fungus. Discs were cut from PDA + mushroom, 4 were placed discs about he beam of the sheet without wounds (Figure 16). HE they made 4 repetitions.

Treatment witness 4. Application of PDA. HE they cut discs of P.D.A. pure without fungus, 4 discs were placed on the upper part of the leaf without wounds (Figure 16). 4 repetitions were done. All the leaves were placed on the cotton layer inside the tray; this HE covered with plastic transparent by 72 hours for favor the infection of the fungus.

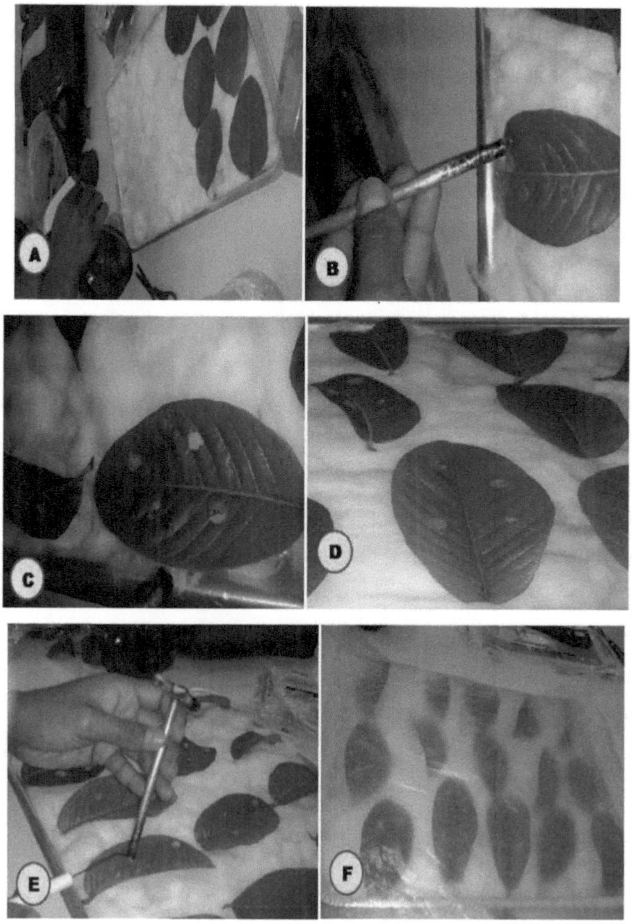

Figure 16. TO) Placement of leaves about cotton in the tray. B) Placement of the P.D.A. in the leaf C) Fungus in the Leaf with punch wounds D) Fungus on the leaf with needle wounds E) Spraying of the inoculum suspension and F) Tray with leaves covered with plastic.

5.7 Trial I. Effect of Trichoderma spp. metabolites against the fungus associated with guava cloves.

Treatments. The treatments evaluated HE describe in he Chart 16 .

Obtaining strains of Trichoderma spp. All species tested in this experiment were prepared in pure culture, using the following methodology:

Isolation and identification of the native strain. In the orchard where the diseased fruits were collected; A sample was also obtained 500 g of soil to insulate the strain native to Trichoderma sp ., with the technique of dilution on plate (box Petri) described by Tuite (1969), which consisted of weighing 1 g of soil and transferring it to a test tube with 9 mL of water distilled sterilized, HE stirred vigorously for homogenize the sample, HE took 1 mL and HE passed to the second tube with 9 mL of water, stirred again, 1 mL was transferred to the third tube, this was also shaken, then an aliquot of 1 mL was taken and distributed with a glass rod over the surface of solidified PDA; were made three repetitions; It was incubated for 4 to 5 days at room temperature. + 28 °C in the laboratory ; The boxes were inspected, greenish colonies with characteristics of Trichoderma were located; of one of them HE they took tips of hyphae (Singleton et al ., 1992) and were transferred to PDA in a Petri dish (Figure 14), to obtain the beneficial fungus in pure culture, which was preserved in PDA in a test tube.

Chart 16 . Treatments used in he Rehearsal Yo

No. of Treatment	Treatment or product commercial	Ingredient asset
T1	Witness	-
T2	Trichoderma sp . (native strain Tetipac)	-
T3	T. fasciculatum Bissett	FITHAN
T4	T. reesel Simmons	, BACTIVA
T5	T. asperellum (onion Chilapa)	Chilapa
T6	Trichoderma sp.	BIOGRAPHY
T7	T. asperellum (onion)	Cucula

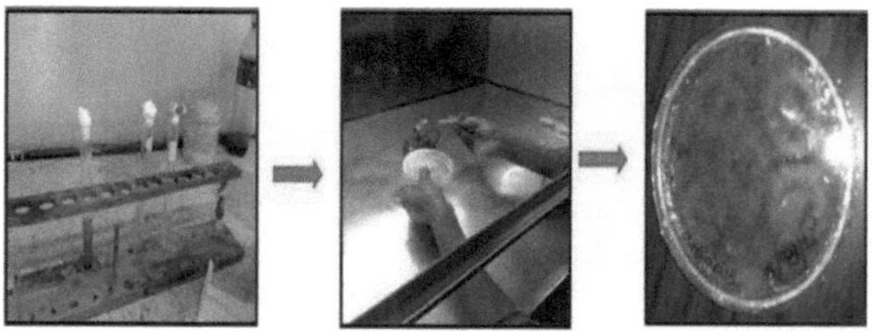

Figure 17. Isolation and purification of the strain native of Trichoderma sp.

To carry out the identification of this native strain, a preparation was made in lactophenol, from the purified fungal colony, the morphological characteristics of mycelium, conidiophores, phialides, conidia and chlamydospores were observed in the compound microscope, and the fungus was identified at the genus level, using the Barnett and Hunter keys. (1998) and Watanabe (2000).

Isolation of commercial strains. Strains from the Natu-Control (T. harzianum strain BK-Th001) and PHC ROOTMATE (T. virens) products were used ; which were activated and purified by means of the technique (dilution on plate) described previously and HE kept in PDA in test tube.

Laboratory strains. Two strains of T. asperellum (Chilapa and Cocula) were used , which are kept in PDA in a refrigerated test tube, in the Phytopathology laboratory. They were reactivated by growing them in PDA.

Preparation of PDA. Previously prepared PDA was liquefied, doses of 20 mL were added per test tube of 50 mL capacity (42 tubes); A cotton lid was placed on this, which was covered with another one made of aluminum foil held with a rubber band. All the tubes were placed in a rack inside the pressure cooker, they were sterilized at 15 pounds/inch ^{2} ; the tubes were emptied. 20 mL of PDA from each tube in the Petri dish; The medium was allowed to cool and solidify to use the cellophane paper technique.

Figure 18 . Unit experimental (box Petri)

Cellophane paper technique. paper was used sweet commercial cellophane; was cut into circles 9 cm in diameter, same as the surface of the Petri dish; was sterilized along with the test tubes, a piece was placed on the PDA surface in the box. On the other hand, pure cultures of Trichoderma spp., 5 days old, were used, from which with a 0.5cm diameter punch, disks of the fungal colony, were placed in the center of the surface of the half in the box; HE incubated 48 hours, HE removed with slowly clamp the paper with the Trichoderma spp colony; with the purpose of keeping diffused in the culture medium, the metabolites secondary produced by 10 cm.2.0 cm.the beneficial mushroom and, immediately, to try the effect of they about he pathogen, in the same box, HE they placed discs of P.D.A. (0.5 cm of diameter) + fungus, HE incubated to room temperature of + 28°C, in the laboratory and the mycelial growth of the pathogen was measured every 24 hours, per treatment.

Design and experimental unit. The seven treatments were distributed on a table in the laboratory, according to the completely randomized experimental design with six repetitions. The experimental unit was a Petri dish of 10 cm in diameter and 2.0 cm in height (Figure 18).

Study variables. To know the effect of the evaluated treatments on the development of the colony of the investigated fungus, the following response variables were measured :

Diameter from the colony of the fungus . HE they made measurements each 24 h, during 10 days (240 hours).Percentage of inhibition of the cologne of the fungus. HE calculation with the equation. % inhibition = (D 1 -D 2) / D 1) 100 Where:

D 1 = diameter of the pathogen colony growing in boxes with PDA (control) D 2 =

diameter of the cologne of the pathogen growing in the box Petri with PDA,where previously HE development Trichoderma spp ., about he paper cellophane, and released enzymes and secondary metabolites into culture medium.

Statistic analysis. In the data of the diameter of the colonies of the fungus where values of zero, They made transformations using the formula : √ + 0.5 (Reyes, 1981); after HE performed he analysis of variance, using he package statistical Statistical Analysis System (SAS, 2014), of agreement to the design experimental completely to the random (Steel and Torrie, 1998) and he model statistical following:

And $_{oh}$ = µ + you $_{i}$ + e $_{j}$

Where:

Y_{ij} = Answer of the jth unit experiment, with he i- thtreatment.

Yo	=	Subscript of treatment
j	=	Subscript of repetition
µ	=	Overall average
you i	=	Effect of the i-th treatment
e j	=	Mistake experimental of the jth repetition of the ith treatment

The variables that showed the significant effect of the treatments were subjected to a multiple comparison test of means using Tukey's method (as; 0.05).

5.8. Rehearsal II. Effect of extracts vegetables against he fungus associated to the nail of The guava

Treatments. HE they evaluated the treatments presented in he Chart 17.

Preparation of PDA with plant extract. The dose of each plant extract was measured and placed at the bottom of the Petri dish; Next, 20 mL of PDA was added, it was shaken gently to homogenize the mixture, and the culture medium was allowed to cool and solidify; The fungus was planted in the center of the box with PDA; It was incubated at room temperature of + 28 ^{O} C in the laboratory and the growth of the fungus was recorded every 24 hours (Figure 16).

Chart 17 . Treatments with extracts vegetables used in he Rehearsal II

No.	Treatment and Tradename	Ingredient asset	Dose*	
			mLL - 1	mL20mL -1 of PDA
1.	Testigo	-	0	0
2.	CINNOIL	Cinnamomuncassia L. y C.zeylanicumL.	5 – 10	0.2
3.	CAPSIOIL	Allium sativumL.y C.zeylanicumL.	5 – 10	0.2
4.	LIPPOIL	Lippiagraveolens Kunth y L. berlanderi Schawer.	5 – 10	0.2
5.	BIOCANELA	C.zeylanicum L	5 – 10	0.2
6	EXTRACT IN NEEM	Azadirachta indicates A.Juss.	10 – 20	0.4
7	EXTRACT DE AJO	Alliumsativum L.	10 – 20	0.4

Note. The fungicides HE they used in the dose recommended by he maker (DEAQ, 2015)

Design and experimental unit. The seven treatments were distributed on a table in the laboratory, the experimental design and the experimental unit were similar to those indicated in Trial I (Figure 18 and 19).

Study variables . To know the effect of organic products on the development of the pathogenic fungus, the same variables described in the Essay I.

Statistic analysis . It was carried out in the same way as in Test I.

Figure19. Distribution of the 7 treatments with6 repetitions in an experimental design completely random.

5.9. Rehearsal III. Effect of fungicides chemicals against him fungus pathogen

Treatments . The treatments indicated in Table 18 were evaluated.

Chart 18. Treatments fungicides chemicals used in he Rehearsal III

No.	Fungicides		Dose	
	commercial number	Ingredient asset	ha*	20mL -1 of P.D.A.
1.	WITNESS	-	-	-
2.	BENOMIL	benomyl	60 – 90 g	0.012 g
3.	CERCOBIN	thiophanate methyl	50- 90 g	0.06 g
4.	BEEF 200	mancozeb	1.5 – 3 kg	0.03 g
5.	CUPRAVIT	sulfate de cobre	1 kg	0.2 g
6.	ZINEB	ethylene Bis dithiocarbamate of zinc	1 kg	0.2 g
7.	CAPTAN	they capture	250 – 350 g	0.05 g

Note. The fungicides HE they used in the dose recommended by he maker (DEAQ, 2015)

Preparation of PDA with fungicide. The dose of each fungicide was weighed and placed at the bottom of the Petri dish; Next, 20 mL of PDA was added, it was shaken gently to homogenize the mixture, and the culture medium was allowed to cool and solidify; the mushroom was planted in the center of the box with the PDA; It was incubated at room temperature of + 28 ^{O} C in the laboratory and the growth of the fungus was recorded every 24 hours.

Design and experimental unit. The seven treatments were distributed on a table in the laboratory (Figure 16), the same design and experimental unit was used as in the Test YO.

Variables of study. HE they measured the themselves variables described in he Rehearsal YO.

Analysis statistical. HE performed as HE Indian in he Rehearsal YO.

VI. RESULTS AND DISCUSSION

6.1 ID of Pestalotia Speg microspore .

Morphological . Of the samples of fruits with lesions necrotic, isolated, purified and identified at the genus level Pestalotiopsis (=Pestalotia) which is characterized by developing mycelium cottony white, which when mature produces abundant black acervules in a random radial arrangement or in concentric circles (Figure 20A), on the surface of PDA. It has conidia with 5 septa, whitish in color, but when the largest cells in the central part age, they become dark. The hyaline conidiophores, irregularly branched, are septate, smooth and short; They produce dark, straight or slightly curved, ellipsoidal or fusiform conidia (Figure 20B and 20C), which have six cells, with the basal and terminal hyaline; the last of these is pointed with two or more hyaline apical appendages. These morphological characteristics are similar to those described for the genus Pestalotia by Barnnett and Hunter (1998). Molecular . HE achievement the amplification of the DNA of fungus with both games of first: STIs and elongation factor, obtaining a size of 550 to 615 base pairs (bp) (Figure 21). When comparing the sequences obtained in the present work with those available in GenBank, 99 similarities were found with previously reported sequences of P. microspora (Chart 19). The sequences based in he gene of the factor of elongation wereprecise in the ID to level of species; this result corroborated the veracity of morphological identification.

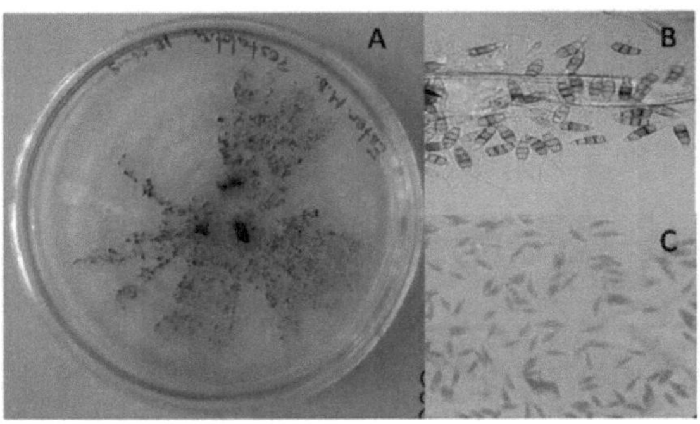

Figure twenty. TO) Cologne of P.microspora in PDA, B) Conidia observed to the microscope 100X, and C) Conidia observed under a 140X microscope.

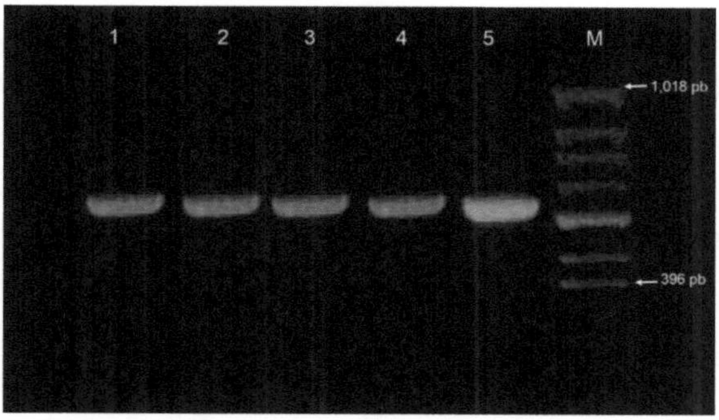

Figure 21. PCR amplification of the elongation factor (lanes 1 to 4) and the ITS region (lane 5) obtained from DNA of P. microspora isolates . M= Molecular weight marker. The arrows point to the molecular weight marker size.

Chart 19. ID by PCR-STI of genes ribosomal of the fungus isolated of fruits of guava (Psidium guajava

L.) based in the comparison of their sequences with the base of data of the national Center for Boitechnology Information®

ID molecular index	of similarity (%) Accession
P. microspora	99 JK436800.1
P. microspora	99 AY924266.1
P. microspora	99 DQ001002
P. microspora	99 HM595547.1

®GenBank2015.

6.2 Proof of pathogenicity (postulates of Koch)

This test allowed us to verify that the purified isolation of P. microspora is pathogenic, because provokes the symptoms typical of anthracnose when HE inoculate in healthy guava fruits and leaves (Figure 22). It was found that the greatest severity of the disease developed in fruits and leaves inoculated with PDA + fungus in tissue with wounds from the punch in fruits and in leaves with wounds from the dissecting needle; likewise, the Symptoms took 10 to 15 days to develop in inoculated leaves without wounds. The fruits and leaves used as control treatments, as expected, remained healthy. The reisolation of the fungus carried out from the tissue artificially infected, presented the same characteristics as the originally inoculated isolate, which fulfilled the protocol of the aforementioned postulates (Bauer, 1984). The symptoms presented on leaves with wounds were similar to those obtained by Keith et al. (2006) who reported that the highest incidence occurred 19 days after inoculation.

Figure 22. Symptoms of anthracnose in fruit and leaves

6.3 ID of the strain native of Trichoderma sp.

From the soil samples, Trichoderma sp . was isolated, purified and identified, whose colonies HE characterize because They present whitish mycelium, growing quickly and, as they mature, they form abundant green air cushions of conidia; supported by hyaline conidiophores, highly branched, not whorled, with individual phialides or grouped in small terminal clusters and sometimes form rings concentric (Figure23). The conidia are ovoid, hyaline and unicellular. The surface of the colony is initially cottony white, over time it begins to sporulate abundantly in the peripheral zone, exhibiting masses of conidia of color white, that then HE they return green grayish. These morphological characteristics observed coincide with those mentioned for Trichoderma sp. by Barnett and Hunter (1998) and Wantanabe (2002).

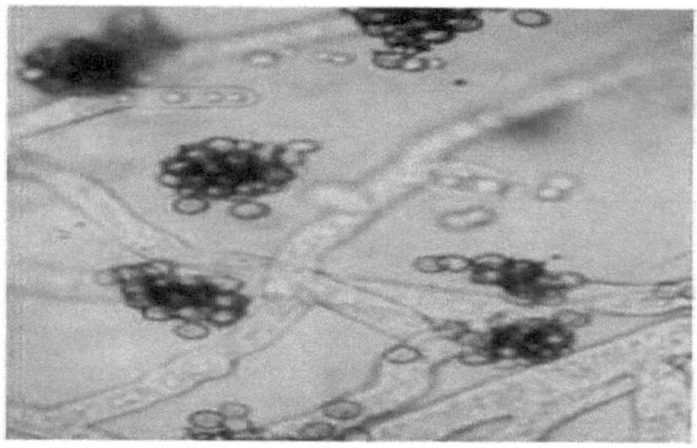

Figure 23. conidiophores, mycelium and spores grouped in masses of the native strain of Trichoderma sp. (100x).

6.4. Trial I: Effect of Trichoderma spp. metabolites against the fungus associated with guava cloves

The data from this test in each of the evaluations are shown in table A-9 a A-18 of the appendix. The study variable presented highly significant evidence ($P<0.0001$) on all measurement dates of the growth of the pathogen colonies (Table twenty). The T. reesei strains and T. asperellum (Strain Cocula), inhibited completely the growth of P.microspora ; on all sampling dates because the metabolites of these two species maintained their greatest effectiveness in inhibiting the pathogen from the beginning of the trial until the last evaluation date, carried out 10 days after the end of the experiment (Table 18). In the control treatment, the fungus grew at a rate of 0.835 cm day $^{-1}$ (Figure 22). At the end of the trial, in the treatments T. asperellum (Chilpa strain), Trichoderma sp., (Tetipac strain), T. fasciculatum and Trichoderma sp., Averages of 6.38, 6.35, 4.30 and 7.35 cm in diameter were obtained from the P. microspora colonies , respectively (Table 18). The species T. asperellum (Cocula strain) and T. reesei presented 100% biological effectiveness (Figure 24). According to reports by Infante et al., (2009), the antibiosis of Trichoderma species is based on the production of toxic metabolites and antibiotics such as gliotoxin, viridin, trichodermin, suzukacillin, alamethicin, dermadin, trichothecenes and trichorzianin; which are responsible for inhibiting the growth of phytopathogenic fungi.

Chart twenty. Comparison of the values averages of the diameter of the colonies of Q. microspore in all the dates of assessment, in he Rehearsal Yo **Date**

No.	Treatment	1	2	3	4	5	6	7	8	9	10
1	Witness	1.15 to	1.7 to	2.2 a	2.7 a	4.1a	4.9 a	5.5 a	6.4a	7.4 a	8.3 a
2	Trichoderma sp. (Cepa Tetipac)	0.95 b	1.4 a	1.8 ab	2.2a	2.7 b	3.2 b	3.7 c	4.4 c	5.1 c	6.4 c
3	T. fasciculatum	0.0c	0.6b	1.0c	1.4b	1.6c	1.9c	1.9d	2.9d	3.6d	4.3 d
4	T. reesei	0.0c	0.0c	0.0d	0.0c	0.0d	0.0d	0.0e	0.0e	0.0e	0.0 e
5	T. asperellum (cepa Chilapa)	0.0c	0.95 b	1.3b c	2.1a b	2.5b c	3.2b	4.1b c	5.1b c	5.9b c	6.4 c
6	Trichoderma. spp	0.95 b	1.4 a	2.0 a	2.5 a	3.2 b	4.1 ab	5.1 ab	5.9 ab	6.4 b	7.4 b
7	T. asperellum (cepa Cocula)	0.0 c	0.0 c	0.0d	0.0 c	0.0d	0.0d	0.0e	0.0e	0.0e	0.0 e

Note. The values averages in the column, with the themselves literal, No are statistically different.

In a research work carried out by Barman et al. (2015) evaluated different strains of Trichoderma, including T. reesei , in the control of Pestalotiopsis theae ; The authors reported that the maximum inhibition in said work was 72.4% with T. viride , the result obtained with the T. reesei strain was 65.2%; The reported inhibition percentages are different and lower than those obtained in the present study where 100% inhibition was recorded with the T. reesei strain.Regarding the T. asperellum strain Csaegro strain, it has been evaluated with isolates of Rhizoctonia solani and Phytophthora capsici under in vivo conditions and it has been observed that it delays the presence of said pathogens between 4 and 6 days, when inoculated in fruits of pipiana squash (Díaz et al., 2014; Díaz et al., 2015). There are extensive reports on the biocontrol of different phytopathogenic fungi through the use of Trichoderma strains. spp., (Barman et al. , 2015; Talla et al., 2015). However, there is little information on Pestalotiopsis microspora .Ortiz (2015) points out that T. reesei, T. fasciculatum and Trichoderma spp. managed to inhibit the growth of the fungal colony, since these contain secondary metabolites that help prevent growth.

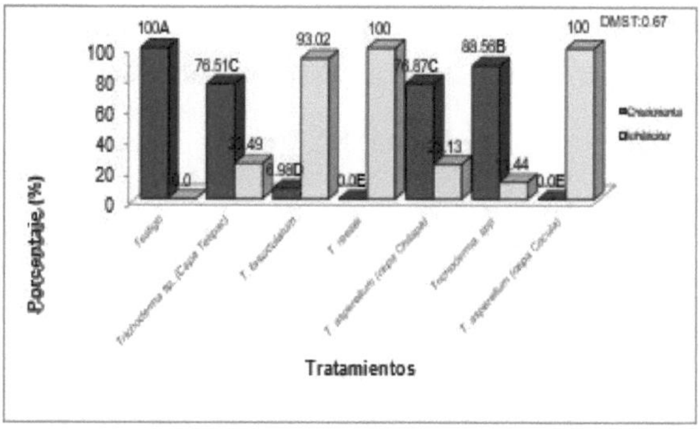

Figure 24 . Percentage of growth and inhibition (effectiveness biological) of the P colonies . microspore , in Trial I. DMST (Tukey's Least Significant Difference)

6.5. Trial II: Effect of plant extracts against the fungus associated with guava cloves .

During the 10 dates of measuring the growth of the colonies of the pathogenic fungus, highly significant differences were found (Tables A-19 to A-28 of the appendix); because the neem extract from date 1 to 10 was the one that most inhibited the development of the pathogen (Table 21). At the end of the experiment, averages are recorded, in increasing order of 6.5, 6.6, 6.7, 7.2 and 7.4 in Biocinnamon, garlic, Lippoil, Capsioil and Cinnoil, respectively(Figure 25).In the experimental units of the control treatment, it was noted that the fungus had an average growth rate of 0.81 cm day $^{-1}$; in such a way that, at 240 h, he managed to completely cover the Petri dish with PDA with a diameter of 8.1 cm. At the end of this trial, it was determined that the extracts tested have a fungistatic effect, because they delay but do not suppress the growth of P. microspora colonies . At the end of the trial, the overall average was 2.58 cm. In a similar trial, Ortiz (2015) compared the biological effectiveness of organic extracts on Pestalotiopsis sp., the same extracts on the same fungus where they got results different in his investigation he extract of Lippoil turned out that achievement

Chart 21.Comparison of the averages of the diameter of the colonies of Q. microspore in 10 datesof assessment, in he Rehearsal II

Fecha* N° Trat.

N°	Trat.	1	2	3	4	5	6	7	8	9	10
1	Testigo	1.2 a	1.5 a	2.6 b	3.3 a	4.0 a	4.9 a	5.6 a	6.4 a	7.4 a	8.1 a
2	Cinnoil	0.5 b	0.7 b	2.8 b	3.4 ab	3.8 a	4.4 a	4.9 a	5.6 a	6.4 a	7.4 a
3	Capsioil	1.2 a	1.5 a	2.6 b	3.3 ab	3.7 a	4.3 a	5.1	5.8 a	6.7 a	7.2 a
4	Lippoil	1.2 a	1.5 a	4.0 a	4.0 a	4.0 a	4.4a	4.9a	5.8a	6.5 a	6.7 a
5	Biocanela	0.0 c	0.5 bc	1.7 c	2.7 b	3.3 a	3.9 a	4.7 a	5.3 a	5.9 a	6.5 a
6	Neem	0.0c	0.0c	0.5d	0.7c	0.9b	1.3b	1.8b	2.3b	2.8b	2.8b
7	Alliioil	0.7 b	1.5 a	3.3 ab	3.7 a	3.7 a	4.2 a	4.8 a	5.6a	6.2	

Note. The values averages in the column, with the themselves literal, No are statistically different, *. The data HE register each 24 h. inhibit the growth of the fungal colony. In the present study, the extracts of neem, cinnamon and garlic recorded the highest inhibition, 35.30, 23.53, 22.36%, derived from a fungistatic action.

These data contrast with Barman et al. (2015), who with neem oil reported 100% in Pestalotiopsis theae ; it found in he present rehearsal differs to it reported by Bhuvaneswari et al. (2010) who with neem extract found a 55.55% inhibition on Pestalotiopsis palmarum . Likewise, with cinnamon extract Bhuvaneswari et al., (2010) found 100% inhibition in P. palmarum , this figure is higher than the inhibition found in this study. The effect of extracts of Cinnamomun sp. has been studied against species of the genus Colletotrichum , it has been reported by Maqbool et al., (2011) and Idris et al ., (2015) and they recommend it as an alternative in the control of diseases caused by phytopathogenic fungi. The garlic extract was also evaluated by Bhuvaneswari et al (2010) and Barman et al., (2015), the first authors cite that with Allium sativum they found an inhibition of 60%, the second authors report a fungistatic effect exerted by the extract of Allium sativum on Pestalotiopsis theae .

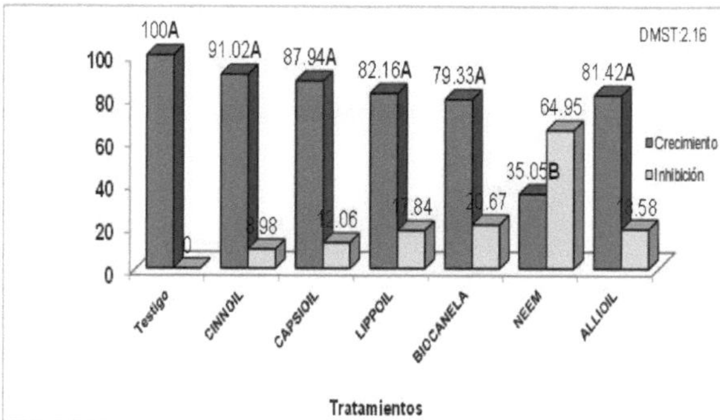

Figure 25. Percentage of growth and inhibition D.E.P. microspore, in the five treatments of Essay II. DMST (Tukey's least significant difference).

Rehearsal III: Effect of fungicides chemicals against he fungus pathogen

The information corresponding to the mycelial growth in each of the samplings carried out by treatment is found in Tables A-29 to A-39 of the appendix; Highly significant differences ($P<0.0001$) were observed on all evaluation dates. In the Manzate and Captan treatments, they exerted fungicidal activity on the pathogen in all evaluations since they did not allow growth, they inhibited P. microspora 100% (Table 22). Cupravit did not allow the growth of P. microspora in the first and second evaluation; In the other evaluations it exerted a fungistatic activity; in benomyl, cercobin treatments and zineb HE they got averages of 6.35, 4.30 and 7.35 cm of growth of the pathogen. In the control treatment the pathogen grew at an average rate of 0.83 cm day $^{-1}$, in a period of 240 h the colonies of P. microspora were 8.5 cm in diameter; value equal to the diameter of the Petri dish with PDA. With the Manzate and Captan fungicides, a biological effectiveness of 100% was obtained; while in the treatments Benomyl, Cercobin, Cupravit and Zineb HE they got 23.50, 48.70 and 11.45 %, of effectiveness, respectively (Figure 26). The results obtained in this investigation demonstrated that he isolation of Q. microspore is highly sensitive to the Manzate and Captan fungicides. Very similar results expressed in % inhibition have been reported by Bhuvaneswari et al (2010) In studies carried out with P. palmarum , those who obtained 100% inhibition with mancozeb. According to Barman et al. (2015), also evaluated fungicides such as Captan and Mancozeb, reported different and lower values than those obtained in this trial, reporting 37.5% with Mancozeb and 82.4% with Captan. The 100% inhibition reported in this work coincides with what was reported Esiegbuya et al (2014), who found complete inhibition with tumban on Pestalotiopsis clavispora .

Cuadro 22. Valores promedios del diámetro de las colonias de *P. microspora* en todas las fechas de evaluación, en el Ensayo III.

N°	Trat.	Fecha*									
		1	2	3	4	5	6	7	8	9	10
1	Testigo	1.2a	1.7a	2.2a	2.6a	4.0a	4.8a	5.4a	6.3a	7.3a	8.3a
2	Benomilo	0.9b	1.4a	1.8ab	2.1b	2.7b	3.1b	3.6c	4.4c	5.0c	6.3c
3	Cercobin	0.0c	0.6b	1.1c	1.3c	1.6c	1.9c	1.9d	2.8d	3.6d	4.3d
4	Manzate	0.0c	0.0c	0.0d	0.0d	0.0d	0.0d	0.0e	0.0e	0.0e	0.0e
5	Cupravit	0.0c	0.9b	1.3bc	2.0bc	2.4bc	3.1bc	4.1bc	5.0bc	5.8bc	6.3c
6	Zineb	0.9b	1.4a	2.0a	2.4b	3.1b	4.1b	5.0ab	5.8ab	6.3b	7.3b
7	Captan	0.0c	0.0c	0.0d	0.0d	0.0d	0.0d	0.0e	0.0e	0.0e	0.0e

Nota. Los valores promedios en la columna, con las mismas literales, no son estadística diferentes,*. Los datos se registran cada 24 h.

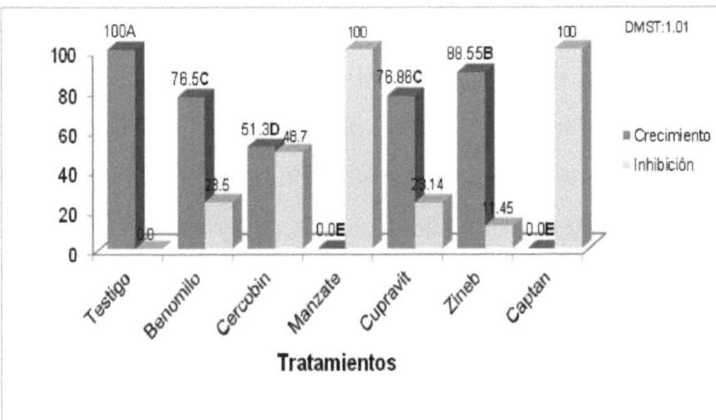

Figure 26. Percentage of growth and inhibition D.E.P. microspore, in the five treatments of the Rehearsal III. DMST (Tukey's least significant difference).

VII. CONCLUSIONS

Of according to the set objectives and in base to the results obtained in the Essays I, II and III, HE concludes the next:

- HE identified morphological and molecularly to the species Pestalotiopsis microspore Speg., agent causal of the nail of guava, in Tetipac, Gro..
- The isolation of this fungus It is pathogenic when inoculated into fruits and healthy leaves guava. The infection was more severe in fruits and leaves with artificial wounds.
- Strains of Trichoderma spp ., present significant differences in inhibition of the pathogenic fungus.
- The species of T. asperellum (strain Cocula), T. fasciculatum, and T. reesel They present 100% biological effectiveness.
- The extracts vegetables show significant differences in the control of the pathogen.
- The extracts Cinnoil, Capsioil, Lippoil, Biocinnamon, Neem and Allioil has fungistatic effects on P. microspora.
- He extract of neem I stand out about the the rest, by have a effectiveness biological of 31.8%.
- The products chemicals Mancozeb sulfate of copper, dithiocarbamate of zinc and they capture, they have action fungicide because they inhibited completely the development of the fungus.
- The products Benomyl and thiophanate methyl they exercise action fungistatic because They only delay the growth of pathogen colonies.

REFERENCES BIBLIOGRAPHICAL

Citrus, NG 1996. Phytopathology. First edition. Editorial Limusa. Mexico D. F. 754 pp.

Citrus, N. g. 2005. plants Pathology. Fifth edition. Elsevier academic Press. Saint Diego,

California, USES. 922 pp. Anonymous.2014c. Morphology Trichoderma spp . Internet edition . http://educacion.ucv.cl/prontus_formacion/site/artic/20070710/asocfile/ASOCFILE1 20070710123449.PDF (Date of consultation: 06/06/2015).

Anonymous. 2015a. Advantages of using Trichoderma . Internet edition. http://www.i nfojardin.com/foro/showthread.php?t=39804 Consultation date: 04/19/15.

Anónimo. 2015b. Antagonistic properties of species – groups of Trichoderma. I. Production of nonvolatile antibiotics. Transaction of the British Mycological Society 57:25–39.

Arévalo, Fr. SAY., Diaz, J. A., Galindo, AJ, Rivero, C. M. 2012. Management phytosanitary of the cultivation of the guava (Psidium guajava L.) Measures for the winter season. pp. 100-1 14-1 Edition de the internet http://www.ica.gov.co/getattachment/00295b79-bcb0-4ab2- 80f9- b6e3ab7218b8/-nbsp;Phytosanitary-management-of-guava-crop.aspx. Date of consultation: 15/06/1

Avalos, c. 2014. He controversial use of agrochemicals. Generation Group. Discharged of the network: http: // www. generation. com / magazine / 876 / controversial - use - agrochemicals. Consulted: 07/15/2015.

Ayala, B. M.F. 2008. Integrated management of moniliasis (Moniliophthora roreri) in cultivation of cocoa (Theobroma cacao L.) through the use of fungicides, combined with cultural work. Bachelor's thesis. Faculty of Engineering in Mechanics and Production Sciences. Litoral Polytechnic Higher School. Guayaquil, Ecuador. 25-32bp.

Barman, H., Aniruddha, R. and Kumar-Das, S. 2015. Evaluation of plant products and antagonistic microbes against grey blight (Pestalotiopsis theae), a devastating pathogen of tea. African Journal Agricultural Research. 9(18):1263-1267.

Barnett, H. and Hunter, B. 2000. Illustrated genera of imperfectfungi. Third Ed.

Burgess Publishing Company. Minneapolis, USA. 241 pp.

Barnett, H., and Hunter, B. 1998.Illustrated Genera of Imperfect Fungi.The American PhytophatologicalD. F. 726 p

Bauer, of the IML 1984. Phytopathology. First edition. Editorial Future, Mexico, D.C. F. 384 pp.

Bhyvaneswari, 2010. In vitro studies on management of leaf blight disease of palmyrah caused by Pestalotiopsis palmarum (Cooke) Ste. Journal of Plantation Crops 38(2): 165-167.

Carrasco, RA 2014. Dictionary of agrochemical specialties. 4th Edition 2014. PLM Mexico SA of CV Mexico, DF 2000 pp.

Chaverry, P. and Samuels G. J. 2003. Multilocus phylogenetic structure within the Trichoderma harzianum/Hypocrealixxi complex.Molecular Phylogenetics and Evolution 27: 302- 313.

DEAQ. 2015. Dictionary of Agrochemical Specialties. Viewed online: http://www.agroquimicos-organicosplm.com/ consulted 06/05/15.

Diaz, N. J.F., Alvarado, g. EITHER .G.,Leyva, M. S. g. Ayvar, S. S., Michael, TO. TO. c. and Vargas,

H. M. 2015. Identification and control of fungi causing fruits rot in pipiana pumpkin (Cucurbita argyrosperma Huber). African Journal Agricultural Research. 10(11):1150-1157. DOI: 10.5897/AJAR2015.9486

Diaz, NJ F., Vargas, HM, Ayvar, S. S., Alvarado, GOG, Solís, A. J. F., Durán, RJA, Díaz, CHL and Hernández, AA 2014. Morphological and PCR identification of Rhizoctonia solani Kühn from pipiana squash fruits and their management in the greenhouse. Biotechnics. 16(3):17-21.

Domingo, FP, Insuasty, O. and Fanor. C. 2006. Spatio-temporal distribution and damage caused by Pestalotia spp. in guava fruits. Body magazine. Edition Internet : http://www.ica.gov.co/getattachment/d08ab9dc-7593-4315- 9a49-b3a5f52d2da2/Publication-20.aspx.Date consultation: 04/19/15.

Esiegbuya, E. A., Oruade-Dimaro, E. E., Odigie, F. I., Okungbowa, O., Lgb Inedion, L. G. B. and A. Ojieabu. 2014. In vitro Evaluation of Some Selected Fungicides against Pestalotiopsis clavispora and Pseudocochliobolus eragrostidis Isolated from

Vitellaria paradoxa Seedlings. Journal of Agriculture and Veterinary Science. 7(1): 80-85.

Ezziyyani, M. 2004. Biocontrol of Pestalotiopsis microspora (Psidium guajava L.) through a combination of antagonistic microorganisms. Murcia: Doctoral Thesis. University of Murcia.

Flowers, Q. YO. 2015. Integrated management of nail in guava (Pestalotiopsis psidii). p 1-19. Consultation date: 06/16/15.

Gams, I. and Bissett, M. 1998. Morphology and identification of Trichoderma in: Trichoderma and Gliocladium (Vol. 1).

Garcia, EM 1973. Modifications to the system of classification climatic of Köppen. Institute of Geography. National Autonomous University of Mexico. Mexico, DF pp 32-41.

Gonzalez, g. AND., Castle, Q. YO. and Lush, M. TO. 2008. Control of the nail of the guava. Memoirs of the XXXI National Congress of Biological Control Society Mexican Biological Control. pp: 446-429.

Guédez, C., Castillo, C., Cañizales, L. and Olivar, R. 2008. Biological control: A tool for sustainable and sustainable development. Internet edition. http://www.saber.ula.ve/bitstream/123456789/29752/1/articulo5.pdf. Consultation date: 06/17/15.

Harman, G.E., Howell, C. R., Viterbo, A., Chet, I. and Lorito, M. 2004. Trichoderma species- opportunistic, avirulent plants symbionts. Microbiogy 2: 43-56.

Hernandez, EITHER., Herrera, R. and Castle, H. 2009. Trichoderma spp ., a alternative for he control of phytopathogenic fungi. Scientific Dissemination Magazine. True Science. Digital edition #17. Downloaded from the net: http: // www.postgradoeinvestigacion.uadec.mx/CienciaCierta/CC17/cc17trichoderma.ht ml. Consulted: 06/10/15.

Idris, FM, Ibrahim, AM and SF Forsido (2015) Essential Oils to Control Colletotrichum musae in vitro and in vivo on Banana Fruits. American-Eurasian Journal of Agricultural & Environmental Sciences 15(3): 291-302.

Infant, D., Martinez B. and Reyes, L. 2009. Trichoderma action mechanisms against phytopathogenic fungi. Rev. Protección Veg. 24 (1): 10 - 84.

Insuasty, B. O; Monroe, R; Díaz, F. A; y Bautista, D. J. 2005. Management phytosanitary del cultivation of guayaba (Psidium guajava L.) in Santander. p 29-30. Consultation closes: 06/15/15.

Jiménez, A. and Santos. R- 1992. Biological and morphological study of the fungus that causes rot apical of the fruits of guava (Psidium guajava L.). Magazine of Agronomy faculty (LIGHT). 9:77-96.

Kasar, P., Maity, S. S., Bhattacharya, A. Y. and D.C. Khatua. 2006. Occurrence of guava fruit canker in west Bengal and bioassay of fungicides against pathogen. Journal of Horticultural.CABAbstracts.http:/ www.cababstractsplus.org/abstracts/Abstracts.aspx?AcNo.=20053079049. Fecha de consulta: 20/06/15.

Keith, L. M., Velasquez, M. E., and Zee, F. T. 2006. Identification and characterization of Pestalotiopsis spp. causing scab disease of guava, Psidium guajava, in Hawaii. PlantDis. 90:16-23.

Kullnig-Gradinger, C. M., Szakacs, G. y Kubicek, C. P. 2002. Phylogeny and evolution of the genus Trichoderma: a multigene approach. Mycological Reseach 106:757-767.

Lopez, F. LC1994. Trial with fungicides and insecticides against "clove" of guava in the Calvillo-cañones region, Colomos Ags.1994-1995. Annual Research Report. SARH-INIFAP-CIRNOC-CEPAB. Mexico. 9 pp.

Maqbool, M., Ali, A., Alderson, P. G., Mohamed, M. T. M., Siddiqui, Y. and N. Zahid. 2011. Postharvest application of gum arabic and essential oils for controlling anthracnose and quality of banana and papaya during cold storage. Postharvest biology and technology 62(1): 71-76.

Martinez, B., Danay, I. and Reyes, Y. 2013. Trichoderma spp. and his function in he control of pests in crops .Internet edition . http://scielo.sld.cu/scielo.php?script=sci_arttext&pid=S1010-27522013000100001. Consultation date: 04/19/15.

Martinez, FM 2012. Evaluation of three strains of Trichoderma spp., as a biocontrol alternative against Phytophthora capsici L. in bell pepper seedlings under a greenhouse. Master's Thesis. Interdisciplinary research center for comprehensive regional development. Institute National Polytechnic. Durango, Mexico.

Martinez, S., Terrazas, E., Álvarez, T., Mamani, O., Vilaa, J. and P. Mollinedo. 2010. In vitro antifungal activity of polar extracts of plants of the genus Baccharis on phytopathogens. Rev. Boliv. Chem. 27:13-18.

Michel, A. TO. c. 2001. Strains natives of Trichoderma spp., (Euascomycetes: Hypocreales), its antibiosis and mycoparasitism on Fusarium subglutinans and F. oxysporum (Hyphomycetes: Hyphales). Doctoral Thesis in Sciences. Tecomán, Colima: University of Colima. 140 pp.

Montes, B. R. 2006. Solid, aqueous and hexane extracts of plants to combat Aspergillus flavus Link in corn. Mexican Journal of Phytopathology. fifteen: 26-30.

Montiel, M. d. R, and reveal, M. J. J. 2001. Etiology of the disease "Nail of The guava". 1-eleven. Date of consultation: 04/19/15.

Ortiz, MG 2015. Investigation report in the phytopathology course I. Center for professional studies. School Superior Agricultural of the State of Warrior. Cocula, Gro. 68-75 pp.

Plancarte, g. Q. J. 2014. Effectiveness biological in vitro of Trichoderma spp. Fungicides chemicals and products organic against Colletotrichum tropical Red and Samuels causing noni anthracnose. Bachelor's thesis. Center for Professional Studies. Higher Agricultural College of the State of Guerrero. Cocula. 15 p.m.

Prakash, O. and B. K. Pandey. 2007. Current scenario of guava diseases in India and their integrated management. In: Proceedings First International guava Symposium. G. Singh, R. Kishum, and E. Chandra (eds.) Acta Horticulturae 735:495-505.

Quiagen. 2012. Manual of the kit QIAamp® DSP DNA Blood Mini. Version 2. Germany.

Quijada., O., and R. Gómez. S. 2005. Technological development for postharvest management of guava in Colombia and Venezuela. Project progress report. National Institute of Agricultural Research of the state of Zulia - Colombian Agricultural Research Corporation. CIMPA Experimental Station (Program National of processes Industrial). Barbossa. http//: www.fotagro.org/projects/01_21_guayaba/III_infotec_01_021.pdf. Consulted: 05/20/15.

Ray, G.O., Quijada, N., Guanipa, R., Carnacho, Y. 2005. Control of apical rot of the fruit and drying of the tree through comprehensive management of the guava crop (Psidium guajava L.). Bioagro 15:135-142.

Rayo, C. c. 2014. Report of research on he course phytopathology I. Center of professional Studies. School Superior Agricultural of the State of Warrior. Cocula, Gro. 68-75 pp.

Kings, TO. J. c. 1981. Design of experiments applied. First Reprint. Thresholds. Mexico,d. F. pp 285-309 .

Romero, RL 1988. Etiology of the "guava clove" disease. XXVIII National Phytopathology Congress. Society Mexican of Phytopathology. p:140.

Rosales, K., E. Sanabria, M., Rodríguez D., Camacaro, M., Ulacio, D., J. Cumana, L. and O. Crescente. 2002. Potential fungicidal effect of plant extracts in he in vitro development of the fungus Colletotrichum gloeosporioides (Penz.) Penz. & Sacc. and anthracnose in fruits of mango. UDO Agricultural Magazine 9 (1): 175-181.

SAGARPA, 2014. Mexico is first in the world in guava production. Consulted online: http://www.siap.gob.mx/produccion-chile -verde/ Date of consultation: 07/02/2015.

SAS Institute Inc. 2014. SASuser's guide: Statistics. Relax 6.03. Ed. SAS Institute incorporation, Cary, NC USA. 1028p.

Singlenton, L. L., Mihail J. D. and Rush Ch. M. 1992. Mehtods for research on soil borne Phytopathogenic Society. Minesota. 4a. ed. 217 p.

Steel, G.D. and H. Torrie, J. 1998. Biostatistics: beginning and procedures. Second Edition.

McGraw Hill.Mexico, d. F. 392 pp.

Thompson PLM. 2012. Dictionary of specialties agrochemicals Edition No. 22. Mexico,D.F. 1840p.

Thomson. 2012. Dictionary in specialties agrochemicals. 22a Edition. Mexico, D. F. 1840pp .

Tuite, J. 1996. Plant pathological methods; fungi and bacteria. Burgess, Minneapolis USA. 239 pp.

Villegas, LOVE 2014. Trichoderma pers. general characteristics and his biological potential in the agriculture sustainable. Orius Biotechnology Magazine. South America. 24 (1): 5-12.

Wantanabe T. 2002. Pictorial Atlas of Soil and Seed Fungi. Morphogies of Cultures Fungi and Key to Species. Second edition. CRC Press. New York Washington, D.C. 500p.

White, T.J., Bruns, T., Lee, S., and Taylor, J. 1990.Amplification and direct sequencing of fungal ribosomal RNA genes for phylogenies. In: M.A. Inns, D.H. Gelfland, J.J. Sninsky, and T.J. White (eds.). PCR Protocols.pp.315-322. Academic Press. San Diego, CA.

MIX
Papier aus verantwortungsvollen Quellen
Paper from responsible sources
FSC® C105338

Printed by Books on Demand GmbH, Norderstedt / Germany